On Privacy Matters:

A Citizen's Guide to Modern Technology

Achutha Raman

A 4PrivacyMatters Publication
New Haven

Cover Design via Canva

Printed in the United States of America
First Edition

ISBN: 978-0999390306
Library of Congress Control Number: 2017914348
LCCN Imprint Name: 4PrivacyMatters

Contents

*****

Preface

We are at a decisive moment that will set the course for the future, near and far. Nourished by the manifold data pipelines of our interconnected world, exciting technologies like artificial intelligence are growing before our eyes. Others, including the Internet, are entering adulthood. Still other technologies once thought well-developed, such as telecommunications, are being thoroughly reconceived. The tools of tomorrow, all of these can provide unbounded value. Yet, they also concentrate immense power that may be wielded for harm.

Beyond physically re-shaping our highways and urban spaces, these technologies will also affect the civil liberties upon which America stands. Central among these liberties is privacy. Although literally absent from the United States Constitution, privacy's spirit has plausibly been championed as substantiating several amendments in the Bill of Rights, including the first, fourth, and fifth. Whether privacy includes a right to be left alone by paparazzi or government operatives, a right to secrecy for intellectual property, a consumer's right to transparency about corporate use of personal data, less, or more, it permeates nearly every facet of public and personal life. Privacy also sits at the heart of difficult conflicts involving national security, free expression, and corporate bottom lines.

There is much to disagree about with these issues and likewise within the pages to come. In displaying discord, this work thus follows from the increasingly contentious tenor of public discourse today. Yet, my hope is that this book will inspire discussion rather than divide. Morality may not be quite as relative as an angst-ridden Hamlet once claimed, but it is true that there are many things neither good nor bad. By and large, technologies themselves are neutral. It is their use that activates duality, and each chapter of this book will explore those opposing possibilities for a particular technology. In each chapter, two different perspectives will be presented, one broadly encouraging development, and one more anxious. Valid points, I hope, can be appreciated from both viewpoints.

Even through their opposition, these arguments should highlight the long common ground. There are many common themes including the value of security and democracy. Most fundamentally, citizen engagement is essential as well. We will need all citizens of our society—corporate executives, government leaders, and private consumers—to work together. We will need to do so with a holistic approach so that we can admire the magnificent manes of these technologies, whilst also acknowledging the fierce talons which extend as threats to privacy and other fundamental rights. We will, moreover, need to engage swiftly, before the opportunity for technological control zooms by.

I hope that this book can help to expand the common ground much further through continued discussion. After all, the technologies discussed affect all members of society. Via online dialogue, book club engagement, and forthcoming coursework, 4PrivacyMatters is committed to furthering an informed discourse about digital privacy topics. Additionally, I hope that in your own family and community groups this book will encourage discussions. Armed with relevant data and an adaptable analytic framework, we as a society can shape the debate in both the near and long term. In turn, we can construct sound policies to employ these technologies as we see fit, rather than being shaped by them. I hope you will join in shaping the future ahead!

Introduction

"Rose...bud..."

It's neither a tree nor a sobriquet for a lost love. It isn't a city and it isn't a song. In a way, it is a lunatic raving, and in another light it is a real place. Certainly, it resonates as a mystery, and in *Citizen Kane*, that peculiar first word hangs over it all: *"Rosebud."* Orson Welles, like his film's leading reporter, Jerry Thompson, invites us to probe Charles Foster Kane's difficult childhood and tortuous if financially successful adult ventures as media mogul and pioneer. Through Thompson and Welles, we learn the sordid details of Kane's business endeavors, romantic affairs, and jealous passions. From the film's start, Welles's camera pushes in to enter Kane's palace of Xanadu, where he has actively sought solitude. From then on, we sit there with him, like a fly on the wall, albeit a fly with spectacular vision.

For filmmaker Welles, his work is a technical and narrative wonder. It has been anointed one of the greatest films of all time, and is among the most influential, having impacted diverse filmmakers including Steven Spielberg, Agnès Varda, Akira Kurosawa, Martin Scorsese, and Steve McQueen. For those of us watching in the audience, the film is a transformative experience, enjoyable if somewhat dated in its theatricality, though certainly not in its spirit of innovation. That mystery it cultivates so well incites us to stay engaged and perhaps develop an overwhelming obsession. We keep watching. We want to learn the truth. For the character of Kane, however, the film's exploration is easily an invasion of privacy.

Kane's life is a troubled, pained, and occasionally deceitful one, though hardly criminal or obviously immoral. Nevertheless, his story is published and broadcast for all the world to see. Perhaps these revelations offer some sort of cosmic justice—after all, the newspapers he owned engaged in an even more salacious sort of journalism, if it could even be so called. His *Inquirers* published candid pictures and speculated on unsavory details. They did so not for public knowledge or any greater good, but simply for the sale of more papers. Circulation his golden idol, Kane forged on, unconcerned with the consequences.

Yet, perhaps this sort of justice is mistaken. Kane may have been a fiction, but the widely known reality was that the film's titular citizen, was in fact tightly based on a true man of flesh and blood: William Randolph Hearst. Some life events were altered, of course, and characters' names pseudonymized, but many of the moments mirrored Hearst's reality. Through the film's release, details that likely never would have achieved notoriety otherwise became public feed.

Still alive at the time of the film's release, Hearst was never consulted for consent or cooperation in the production. On the basis of various civil grievances, he even attempted to sue Welles and block the film's distribution, though the suits were ultimately unsuccessful.[1] Instead, we are left with the naked exposure of one man's life, and another's sculpture of exquisite art.

Privacy and Technology

The term *privacy* has seen scores of definitions through time, and it hangs as a loose, ragged sail over many rafters. Its first legal formulation in America belongs to Louis Brandeis and Samuel Warren, who, as precocious youths in 1890, defined it as the "right to be let alone."[2] Theirs was a direct response to the sort of salacious journalism Kane both suffered and produced, since Warren's wife, a woman of high Victorian society, could not escape the leering gaze of the tabloid press.

Common usage and subsequent legal employment of privacy in America have expanded further. In the 1950s, after decades of the term's unfettered use, Berkeley law professor William Prosser published an

---

[1] The series of lawsuits that Hearst and his descendants launched against Welles and the film's distribution all fell fruitlessly. See, for instance, History.com, "William Randolph Hearst Stops *Citizen Kane* Ads."

[2] Warren and Brandeis, "The Right to Privacy."

influential accounting of the torts, or legal species, of privacy.[3] Fourfold, these were (1) intrusion upon seclusion, (2) public disclosure of embarrassing private facts, (3) false portrayal of a subject in the public eye, and (4) appropriation of another's name or likeness for personal advantage. Ranging from solitude to exposure to libel to identity theft, these ideas are hardly congruous.

Nonetheless, Prosser's designation came to be accepted as the way things had become, if not how they ought to be. His classification lives on in American law, though many other thinkers have offered alternative articulations of idealized privacy.[4] Moreover, in common parlance, *privacy* remains an ever-flowing notion, from "private property" to "private parts" to "Private Ryan" to "privacy of thought." It can even be applied to different perspectives that trigger very different results even within the same case—individuals, society, and corporations all arguably have privacy rights. Naturally, an individual and an incorporated body—whether society or corporation—are not always aligned. Despite this imprecision, privacy's core nonetheless remains of prime importance.

Since the turn of the twentieth century, the emotional and physical harm caused by revelation of private facts has had devastating effects, endangering undercover operatives abroad and driving individuals to suicide. On the other side, privacy in the sense of secrecy has allowed the NSA to collect with impunity both metadata and social media message content from American citizens.[5] Even as corporations are beginning to loudly broadcast their privacy policies that protect consumer data, the same notion of privacy has also allowed corporations

---

[3] Prosser, "Privacy," 389.

[4] See, in particular, Caroline Kennedy and Ellen Alderman's *The Right to Privacy* as well as Neil Richards's *Intellectual Privacy: Rethinking Civil Liberties in the Digital Age* for holistic reappraisals.

[5] See, for instance, Szoldra, "This is Everything Edward Snowden Revealed."

to obscure the precise methods by which they use and sell private information. Attacks on privacy can hence have both minor and major effects, depending on how they are framed.

While enshrined in various European constitutions as well as in the United Nations Declaration of Human Rights, privacy is not an absolute. There are competing concerns that can at least at times overwhelm privacy ones. In the United States, several of these competing provisions are encoded in law, allowing police officers, for instance, to enter a private abode upon noticing "suspicious activity" and allowing comics as well as satirists to present blatantly false information about an individual for the sake of humor and perhaps deeper social commentary. Fundamentally, privacy exists in some tension with the First Amendment's right to free expression: while the right to private thought may justify free expression, news articles, gossip, and idle speculation can deeply intrude into others' private spheres. An individual's right to privacy also conflicts with national security interests that may require government surveillance of private messages for the sake of securing the body politic. Moreover, many other challenges emerge when considering different vantage points—whether the individual, society, or a corporation. These conflicts as well as others will all be discussed in more detail within this book.

This work examines emerging technologies including algorithms, drones, personal genomics, and the Internet of Things through the lens of privacy. With all the rival definitions floating around, none has clear dominion over the rest, and this book will not attempt to explicitly define *privacy*. Nonetheless, a critical element of privacy will be assumed here. Namely, that privacy involves a choice: the choice of individuals to value—or not value—particular elements of their person. My hope is that our commonsense proclivities will help fill in the rest where needed so that the discussion can turn to matters of more immediate practical import.

Another core notion is likewise left undefined: *technology*. Broadly, as used here, the term includes both physical products (such as drones) and computerized techniques (such as algorithm use and content deletion on the Internet). A thing that has both captivated and appalled, technology inherently conveys power. This includes both the power to nourish, and the power to destroy. Perhaps one of the most vocal critics of technology focused on another possible detriment of technology: its ability to fast-track civilization and move things along at "inhuman" rates. In Henry David Thoreau's view, this came at the expense of appreciating the world around us, and of genuinely interacting with it. One of his solutions was technological seclusion, ensconcing himself in the woods near Walden Pond and looking inside of himself.

Today, technology has certainly enhanced that "inhuman" rate. Online web chats transfer communications far faster than the telegraphs of Thoreau's day. Drones can enter narrower spaces than even the tabloid chasers around Hearst ever could, while wielding cameras of far superior vision. In fact, the "inhuman" is itself advancing, as artificial intelligence grows whether with IBM's Watson, Apple's Siri, or Sophia, the world's first AI citizen. Yet, this growth also indicates the relativity of what seems "natural." Other elements of technology have had the opposite effect of division—the electron microscope, for instance, has allowed scientists to study the structure of plants and uncover the complex majesty of their genetic codes. Only through technology could we have uncovered as much and come to appreciate as we do, those very same woods in which Thoreau once took shelter.

The power wielded by technology means that while it poses stark risks to privacy as well as other civil liberties, it also harbors immense benefits. Both effects of these future tools will have to be explored, and in the coming pages we will do so.

Methodology

This work will explore issues related to privacy and emerging technologies using public sources. These public sources, carefully vetted for corroboration, include books, news articles, journal articles, and corporate reports. Rather than relying on exclusive personal interviews, this work is a research effort that makes use of perspectives and resources already available. In turn, I hope this will allow for transparency and repeatability of analysis.

As the present is already trending, our world is becoming ever more interconnected and data traverses national boundaries more easily, rapidly, and perhaps less securely than ever before. Nevertheless, this book will restrict its scope. Predominantly, the gaze will be upon developments in the United States of America, though inspiration is inevitably taken from developments abroad.

The differing perspectives under discussion do not always inhabit the extremes, nor do they portray all the intriguing combinations of beliefs possible on the topic. However, in conjunction they do strive to highlight key arguments on divergent sides of the issue at hand. Perhaps of note, even while the sides are fundamentally opposed on a central issue, readers are likely to find that the perspectives in fact share much in common. This common ground may well be a starting point for further engagement and deeper discussion.

Structure and Content

The first chapter offers a brief timeline of American privacy relevant to the issues discussed later in the book. Each subsequent chapter focuses on a particular technology. For each of these, key questions and a short case open the section. Though some possible answers to the questions will be tacitly offered through the chapters, each of these queries is essentially an open one: there is no decisive right and wrong. The short case, a mini-story, will illustrate some of the elements relevant to the topic under discussion. However, each of the chosen topics covers quite a vast area, and no scenario can adequately represent the whole. The cases are meant to approximate, as much as possible, key considerations necessary for each technology.

Every chapter, save the first, will include a brief introduction to the technology under consideration, elucidation of two perspectives—for and against some particular aspect of the discussed technology, and then a brief conclusion. Scattered through this discussion are a few boxed "spotlight" sections, highlighting related topics not explicitly discussed in the rest of the chapter. Some sources for further exploration are featured at the end of each chapter. At the close of the book, you will also find a full list of consulted sources. Although this produces some redundancy, my aim is to highlight these references for easy exploration.

Digital privacy, the topic of this work, is enormous, especially since the United States has an idiosyncratic privacy regime. My hope is that this work will offer a broad tapestry that covers important areas, even if leaving some gaps. To that end, this book will jump from topic to topic, and hence chapters need not be read in order. The first chapter alone is unique, highlighting in brief a few key events in American privacy relevant to this book. The rest of the chapters will tackle particular technologies and emergent privacy considerations.

Throughout the work, various concepts and issues will repeatedly arise. The skeleton of arguments below can be a tool for framing opposition or support for the various technologies in general:

- General Arguments
    - The value of transparency
    - The value of secrecy
    - The fruits and toxins of interconnectivity
- Socio-political Arguments
    - The defense of democracy
    - The defense of national security interests
    - The empowerment of citizens
    - The protection of citizens from abuse
    - Citizen responsibility and the importance of citizen activity
- Economic Arguments
    - The value of efficiency
    - The importance of acquiring financial profits
    - The importance of maintaining a free market
    - The value of competitive balance

Through the eight technologies discussed, a picture of privacy as a whole should emerge, and I hope that the discussion will allow you to deepen your own understanding and engage with a diversity of perspectives. Let's dive in!

Chapter 1: A Brief History of American Privacy

Several volumes could easily be written on the history of privacy alone. This short timeline restricts its scope to major events in American digital privacy that are particularly pertinent to the chapters that follow.

1620: The Pilgrims arrive in New England and establish **Plymouth Colony**. Fleeing England, the Puritan immigrants seek the freedom to practice their religion without fear of persecution. Like the earlier Jamestown colonists, some also desire new economic opportunities. Although still British subjects, the Puritans of the New World, living more than three thousand miles and an ocean away from their mother country, craft their own customs and government. These include "town halls" at which community members voice concerns and priorities for action. Thousands of other immigrants soon follow the Pilgrims, some by choice, others in bondage. Even in these early colonies, a wide group of people speaking a wide range of languages mixes, though not always amicably.

1762: Jean-Jacques Rousseau publishes **"The Social Contract,"** an essay theorizing the ideal political community. Enlightenment literature, including work by John Locke and Montesquieu, proves highly influential to the American founders. Thomas Jefferson, James Madison, Alexander Hamilton, John Jay, and others incorporate notions of a "right of revolution" and inalienable human rights such as "life, liberty, and property" to craft their own foundational doctrines in the Declaration of Independence (1776) and United States Constitution (1788). With these documents, the "great American experiment" begins.

1791: The original ten amendments to the US Constitution are ratified. Notably, the **Bill of Rights** includes a right to free expression (First Amendment), prohibition of unreasonable searches and seizures (Fourth), rights to due process of law and legal defense (Fifth and Sixth), and an amendment asserting that other rights not explicitly enumerated by the document may still exist (Ninth). These amendments form a basis for deriving privacy law in the future.

1798: Concerned about forces antagonistic towards his presidency and dangers to the nascent republic, President John Adams signs the **Alien and Sedition Acts**. Under the controversial legislation, the government has broad discretion to put on trial and imprison critics. Widely viewed as assaults upon the First Amendment, the acts are allowed to expire in 1801 after ferocious dissent by Thomas Jefferson and other advocates of free expression.

1803: The US Supreme Court case ***Marbury v. Madison***, presided over by Chief Justice John Marshall, introduces and cements the idea of judicial review. Via review, higher courts have the power to examine and overrule the decisions of lower courts, with the Supreme Court providing the final judicial decision whenever it so chooses.

1863: During the American Civil War, Congress passes the **Habeas Corpus Suspension Act**, legitimizing a policy implemented by President Lincoln. With suspension of habeas corpus, captured prisoners can be indefinitely imprisoned. This avoids public trials that might galvanize the Union's enemies and saves time and expenses from the legal proceedings. However, the suspension of civil liberties, even in wartime, was and remains controversial. During the Second World War, President Franklin Delano Roosevelt would implement a similar policy in authorizing mandatory internment camps for hundreds of thousands of Japanese Americans. The government argued that without their incarceration, national security would be endangered. As a result of the policy, thousands die under subpar living conditions, including scores of American citizens.

1873: One of the first anti-obscenity movements in the United States sees passage of the **Comstock Law**. It follows from various state anti-obscenity laws already in place, and bans use of the federal postal system to distribute such items such as erotica, contraceptives, and personal letters alluding to sexual content. Many states develop stricter laws prohibiting sale and possession of these items. Within the next century, however, further legislation and court rulings—notably *Griswold v. Connecticut* (1965), targeted at contraceptive sales—undermine Comstock on the basis of a right to privacy.

May 1879: The modern **California State Constitution** is ratified. One of the longest constitutions in the world, it has hence been cited as both a prime example of what a constitution should not be, and alternatively as a marvelous protector of civil liberties. Article I, Section I explicitly acknowledges Californian residents' right to privacy: "All people are by nature free and independent and have inalienable rights. Among these are enjoying and defending life and liberty, acquiring, possessing, and protecting property, and pursuing and obtaining safety, happiness, and privacy."

1886: In ***Boyd v. United States,*** the Supreme Court unanimously finds that "unreasonable searches and seizures" prohibited by the Fourth Amendment need not only be of one's home, potentially providing the basis for protecting digital and other sorts of invasions.

1890: Samuel Brandeis and Louis Warren publish their influential article **"The Right to Privacy"** in the *Harvard Law Review*. The article articulates privacy as "the right to be let alone," and is a response to a pesky but fascinating new technology of the day: the handheld camera.

September 26, 1914: President Woodrow Wilson signs the **FTC Act** into law. Originally intended to establish consumer protection by breaking up economic trusts, the act creates the Federal Trade Commission (FTC) which will become the primary privacy-protecting body for consumers in the United States.

1917: **The Espionage Act**—enthusiastically upheld in *Abrams v. United States* (1919) and still in effect—prohibits interference with military operations or recruitment during wartime. The act has continued to be enforced, such as with Daniel Ellsberg, publisher of the Pentagon Papers, and again with Edward Snowden. Today, the United States technically remains in a state of war following the 2001 Authorization for Use of Military Force (AUMF).

1932: Aldous Huxley publishes ***Brave New World***, the title ironically seized from Shakespeare's Tempest. Along with Orwell's *1984*, the work remains a giant in dystopian literature. It sketches a future in which individuals have pleasured themselves into submission and become utterly reliant on technology, indifferent to the real world around them.

1934: President Franklin Delano Roosevelt signs **the Communications Act** into law. It mandates the laying of telephone wire and building of radio towers across the country to create a national communications system. Besides the interconnectivity it continues to provide, the law also created new jobs to help the millions struggling in the United States during the Great Depression, and bolstered national security. The act will reign until the 1996 Telecommunications Act upgrades the telecom system.

1935: Originally intended to assist those unable to work in the wake of the Great Depression, **the Social Security Act** has since become a more controversial bill. The act now provides income to retirees and has inspired the creation of a federal database with a unique identifying number for every American.

1936: ***United States v. Curtiss-Wright***, settled by the Supreme Court, establishes that the president has "plenary powers" for foreign affairs.

1941: After a troubled production and greatly antagonistic publicity, Orson Welles releases his début film, ***Citizen Kane***. A massive technical and artistic feat, *Kane* tells the tale of Charles Foster Kane, a man with an unhappy childhood and wealthy but unfulfilling life. Closely modeled on real-life news magnate William Randolph Hearst, who helped usher in so-called yellow journalism, *Kane* is not initially a commercial success. However, for its audacity and craftsmanship, it is now considered one of the greatest and most influential films of all time.

1947: Intending to create a more organized security infrastructure during the Cold War, Congress passes the **National Security Act**. This act designates the structure of military departments in the president's cabinet and also establishes the National Security Council (precursor to the NSA) and Central Intelligence Agency, the United States' first non-military peacetime intelligence agency.

1949: George Orwell publishes ***1984***, which illustrates a dystopian future ruled by a totalitarian government and the omniscient Big Brother. Orwell's work sketches a future to be feared, with ceaseless government surveillance through television sets and public cameras.

1960: In his article **"Privacy,"** legal scholar **William L. Prosser** extracts four disparate torts that have fallen under the name of privacy in prior case law. These torts are (1) intrusion upon seclusion, (2) public disclosure of embarrassing private facts, (3) publicity that places a person in a false light under the public eye, and (4) appropriation of name or likeness. They continue to be the primary way in which privacy cases are pursued in the United States.

1964: The landmark **Civil Rights Act** is signed into law. It broadly prohibits discrimination on the basis of race, color, religion, sex, or national origin. Among other provisions, it makes employment discrimination on any of these factors illegal and prohibits employers from even asking for particular types of sensitive personal information.

1967: In ***Katz v. United States,*** the Supreme Court places substantial restrictions upon wiretaps. Moreover, the case introduces the now-influential "reasonable expectation of privacy" test, whereby privacy torts can be successfully applied only if they occur in situations where a reasonable person would both expect there to be privacy, and where that person actually has an expectation of privacy.

April 1968: Stanley Kubrick launches ***2001: A Space Odyssey***, simultaneously published as a novel by Arthur C. Clarke. The film is another giant of American cinema and uncannily predicts many future technologies, including satellite telecommunication, interplanetary travel, and artificial intelligence, all of which are fundamental to the modern world.

June 1968: The Omnibus Crime Control and Safe Streets Act is passed, including Title III, or **the Federal Wiretap Statute**. This statute makes "wire" and "oral" communications highly protected, with law enforcement required to show "probable cause" before pursuing wiretaps of civilians, and to show that alternative means of acquiring information have been exhausted. These are some of the more difficult clauses to satisfy.

1969: ***Tinker v. Des Moines Independent Community School District*** is settled by the Supreme Court, 7-2. The court rules that students have freedom of expression in school, allowing them to wear black armbands in protest. To legally quell such expression, a school would have to show that the conduct "materially and substantially interfere[s]" with school operation.

October 26, 1970: The **Fair Credit Reporting Act (FCRA)** is approved by Congress. The act establishes a series of rules that credit reporting agencies must follow to increase transparency to consumers under penalty of civil punishment. It is the first major privacy-enforcement law in the United States.

1971: In ***New York Times Co. v. United States***, aka "The Pentagon Papers Case," the Supreme Court rules 6-3 in favor of the journal. Under former President Lyndon B. Johnson, contractor Daniel Ellsberg had leaked documents concerning America's political and military involvement in Vietnam, documents that showed the administration had been lying to the public as well as Congress. The Supreme Court rules that to prevent publication the government would have to demonstrate "serious, imminent harm." At least in this case, national security interests did not overcome the paper's right to free expression in publishing the documents. However, despite the court majority, the ruling was sharply divided, with each justice submitting his own reasoning for the decision, underscoring the difficulty in balancing national security and transparency.

August 9, 1974: **President Nixon resigns from office** with an impeachment verdict in the Senate impending. Nixon was investigated chiefly for authorizing a burglary and illegal wiretap at the Watergate Complex, the Democratic headquarters during the 1972 election. A subsequent cover-up effort dug the president into a deeper hole, exposed by journalists Bob Woodward and Carl Bernstein. His successor, Gerald Ford, would issue a full and unconditional pardon for the former president, citing a need for the nation to heal and move on into a brighter future. Historians and privacy scholars continue to point to President Nixon's abuse of power as a reason to keenly check government surveillance, even when purportedly for national security.

1974: The **Family Education Rights and Privacy Act (FERPA)** is implemented, establishing data privacy rights for students and parents at federally funded schools.

December 1974: President Ford signs the **Privacy Act**, which restricts collection of personal records by the government. It requires that in some cases the Federal Register provide public notice and allow citizens to correct their information. In later years, this act sees various updates, including the Computer Matching Protection Act (1988), which restricts the type and amount of citizen private data that can be aggregated and linked by government employees.

September 1980: The **Organization for Economic Co-Operation and Development (OECD)** accepts **eight principles for data privacy**. Member countries include Australia, Japan, Switzerland, Turkey, the United States, and many of the European Union nations. These nonbinding guidelines are: (1) Collection Limitation, (2) Data Quality, (3) Purpose Specification, (4) Use Limitation, (5) Security Safeguards, (6) Openness, (7) Individual Participation, and (8) Accountability. These principles will continue to be revisited and expanded upon, but the core principles remain. They will also serve as the basis for EU legislation including the 1995 Data Protection Directive and 2016 GDPR.

October 1980: The US **Privacy Protection Act** generally prohibits government employees from searching for or seizing "documentary materials" from journalists. Nevertheless, some exceptions exist, such as when there is reason to believe that immediate seizure is necessary to prevent "the death of, or serious bodily injury to, a human being."

1986: The **Electronic Communications Privacy Act (ECPA)** is signed into law by President Reagan, an amendment to the Wiretap Statute. It expands protections of communications to "electronic communications" but also provides an exception in the case that at least one party consents to surveillance. Some states subsequently implement stronger wiretap laws, such as California, Montana, and Massachusetts, which require two-part consent. Another part of the ECPA (the **Stored Communications Act**) protects stored communications, though does so more weakly.

1988: The **Clinical Laboratory Improvement Amendments (CLIA)** establish standards for clinical lab testing on humans. They are intended to provide basic standards of decency for human test subjects, while also imbuing such tests with greater scientific value.

February 1988: The case ***Hustler Magazine v. Falwell*** is decided by the Supreme Court. The judges unanimously limit the damages a public figure can collect from "intentional infliction of emotional distress."

November 1988: **The Video Privacy Protection Act (VPPA)** is approved following congressional dismay at publication of the video rental records of Supreme Court nominee Robert Bork. Continuing protections from the 1984 Cable Privacy Act, which required that cable companies only collect the necessary information to provide cable services or detect unauthorized reception, the VPPA protects against "wrongful disclosure" of rental or viewing history by video tape service providers. These rules continue to apply to broadband streamers such as Netflix and Hulu. However, an amendment signed into law in 2013 permits consumers to waive these protections.

1989: The **Whistleblower Protection Act** protects employees who reveal misconduct in federal agencies from retaliatory behavior by their employers. However, it hangs in some contention with the Espionage Act.

June 1990: A divided Supreme Court rules 5-4 in ***Cruzan by Cruzan v. Director, Missouri Department of Health.*** The ruling establishes that while most individuals have the right to refuse medical treatment under due process, incompetent persons do not, unless there is "clear and convincing evidence" of their desire. This inspires an upsurge in living wills and other documents explicitly detailing provisions for end-of-life care.

July 1990: The **Americans with Disabilities Act (ADA)** becomes law, prohibiting discrimination on the basis of disability. It particularly impacts employment and workplace conditions.

1993: Named for former press secretary James Brady, victim of a gunshot intended for President Reagan, **the Brady Bill** is signed into law by President Clinton. Among other provisions, the law requires a five-day waiting period for handgun purchases and creates the National Instant Criminal Background Check System (NICS), a federal database now widely used for criminal background checks of volunteers and job applicants.

1994: The **Communications Assistance for Law Enforcement Act (CALEA)** requires that telecom carriers and manufacturers equip their devices and services for targeted surveillance.

1995: The **Telemarketing Sales Rule (TSR)** is first written out, with enforcement to follow from the FTC. Intended to offer users greater choice when it comes to telemarketing, the rule eventually leads to the National Do Not Call Registry established in 2003, which creates a Do Not Call list that consumers can join. With a few exceptions for nonprofits and companies with which consumers have an "existing business relationship," telemarketers must abide by the national registry or else risk substantial fines.

December 13, 1995: The **EU's Data Protection Directive** (**Directive 95/46/EC)** is approved, dictating a first draft of comprehensive rules for data use, data sharing, and data destruction. These rules will be overhauled twenty-one years later in the GDPR.

1996: The **Health Insurance Portability and Accountability Act** (**HIPAA**) is signed into law. As the FCRA established consumer protection laws in credit reporting, this act serves to preserve privacy in the health industry. Primarily intended to streamline American health care and shift it to an electronic system, HIPAA also establishes the basic opt-in system for health care, where the sharing of personal health information requires explicit affirmative consent by the data subject. In 2009, HIPAA will be amended with HITECH, which seeks, among other goals, to incentivize further motion to an electronic health care system.

October 21, 1998: The **Children's Online Privacy Protection Act (COPPA)** is signed into law. The act provides protections for people below the age of thirteen, requiring certain advertisers and website operators to receive opt-in consent for data collection from guardians. In the years since, COPPA has been upgraded by the FCC so that phone apps geared toward children must also comply with the privacy regulations.

October 28, 1998: President Clinton signs the **Digital Millennium Copyright Act (DMCA)** into law, protecting American copyrights and establishing clearer mechanisms to protect intellectual property.

June 2001: In ***Kyllo v. United States***, the US Supreme Court rules that thermal imaging of a home intrudes upon citizens' reasonable expectation of privacy.

September 2001: In the immediate wake of the September 11th attacks, Congress signs the **Authorization for Use of Military Force (AUMF)**. This resolution codifies expanded powers for the president against the al-Qaeda threat. Since al-Qaeda is neither a nation state nor a recognized network, the authorization arguably provides the president with unprecedented leeway, greatly troubling civil liberties organizations.

October 2001: The **USA Patriot Act** specifically expands use of National Security Letters (NSLs) by law enforcement. These letters do not require judicial review and give law enforcement officials an alternative means of investigating citizens.

2002: In a **consent decree** (i.e., a settlement) **with Eli Lilly**, the FTC requires that the company plan and implement a security program to protect consumer data. This is the first time that the FTC explicitly makes such a requirement, but it has since become commonplace in FTC action.

2003: CAN-SPAM prohibits certain kinds of e-mail solicitation, similar to the Telemarketing Sales Rule (TSR). It applies as well to mobile device messages.

May 2005: The REAL ID Act establishes federal standards for official state identity cards. Ostensibly intended for greater security against terrorists, the legislation has been highly controversial, with some fearing that the system severely endangers privacy. Still, states are on schedule to comply with the law by 2020.

December 2005: The **Asia-Pacific Economic Corporation (APEC)** releases its **privacy framework** for "personal information controllers" operating in member countries including Australia, China, Russia, South Korea, the United States, and Vietnam. These nine principles have some divergence from the OECD guidelines: (1) Preventing Harm, (2) Notice, (3) Collection Limitations, (4) Uses of Personal Information, (5) Choice, (6) Integrity of Personal Information, (7) Security Safeguards, (8) Access and Correction, (9) Accountability. APEC endorses these principles as a means to "avoid barriers to information flows, ensure continued trade, and economic growth in the APEC region." In turn, it intends for the framework to benefit consumers, businesses, and governments alike.

2007: Apple releases the **iPhone**, beginning the smartphone and Thing revolution.

May 2008: The **Genetic Information Nondiscrimination Act (GINA)** comes into effect, prohibiting discrimination on the basis of unsubstantiated genetic testing for employment and health insurance.

July 2008: The **FISA Amendments Act** add on to the original 1978 Foreign Intelligence Surveillance Act (FISA). They grant further broad power to national security agencies in investigating individuals, adding on to provisions under the USA Patriot Act. Through FISA, law enforcement is granted an alternative means of investigating citizens, beyond the regular court system. The US Justice Department presents its side of a case to a secret FISA court, which then chooses to grant the wiretap or other request under reduced standards of approval. Section 702 of the 2008 amendments specifically authorizes warrantless surveillance of foreigners and correspondents over 1-year periods. Though the amendments prohibit American communications from being the focus of surveillance, they can be collected incidentally. Section 702 is renewed in January 2018, through 2023.

January 2010: In ***Citizens United v. FEC***, the Supreme Court rules 5-4 that corporate funding of political candidate broadcasts cannot be limited. Several different decisions are authored by the judges, but the majority agree that election funding is a form of speech and under the First Amendment cannot reasonably be limited.

2012: The FTC releases its **report "Protecting Consumer Privacy in an Era of Rapid Change"** (also known as the "Privacy Report") featuring a bevy of recommendations to improve the state of consumer privacy in the United States. The report promotes three major principles for the new ecosystem: (1) Privacy by Design (i.e., incorporating privacy concerns into the cybersecurity of a system), (2) Simplified Consumer Choice (i.e., helping consumers stay informed through simpler privacy notices), and (3) Transparency. Additionally, the FTC floats the idea of do-not-track laws to allow consumers the option to escape tracking by data brokers and other private agencies.

2013: Edward **Snowden** begins his public **disclosures** of classified NSA documents. The disclosures reveal thorough and systematic surveillance by the national security organization, including surveillance of American citizens through electronic devices, and surveillance of internationals. The news comes as a shock to most citizens and even elected representatives, though many also claim Snowden to have been a traitor for his actions.

2014: ***Google Spain v. AEPD and Mario Costeja Gonzalez*** is resolved in the EU's highest court. It is the world's first official formulation of the "right to be forgotten."

February 15, 2015: President Obama releases the memo **"Promoting Economic Competitiveness While Safeguarding Privacy, Civil Rights, and Civil Liberties in Domestic Use of Unmanned Aircraft Systems."** Welcoming UAV development, the memo requires privacy protection via focus on authorized data collection and use, limited data retention, and limited data dissemination. It also recommends broadly protecting the civil liberties of citizens and promoting transparency as well as accountability of drone-using federal agencies.

April 27, 2016: The **General Data Protection Directive** (GDPR: Regulation (EU) 2016/679) is passed. It updates the 1995 DPR with expanded rules covering how data can be transferred beyond the EU and increased control granted to EU residents. It will be enforceable beginning May 25, 2018.

June 2016: The **Federal Aviation Administration** (FAA) releases **guidelines** for domestic drone use by hobbyists and commercial entities. Although there is a focus on safety issues, the guidelines do not aspire to regulate privacy issues. As the document notes, "the FAA's rulemaking authority neither mandates nor permits the FAA to issue or enforce regulations specifically aimed at protecting privacy interests between third parties." Subsequent legislation by Congress, including the National Defense Authorization Act for the 2018 fiscal year, encourages corporate and government drone use.

April 2017: **Joint Resolution 34** is signed into law by President Trump. It undoes various protections that had been advocated by the FCC in late 2016, though were never put into effect. Perhaps most notably, the new bill explicitly removes any restrictions on Internet service providers regarding use of consumer data. For their part, leaders of the FTC and sister regulatory organization FCC (Federal Communications Commission) claim the move will allow American privacy law to develop in a more organized way, centralized at the FTC.

May 2017: The **WannaCry attack** affects three hundred thousand machines across the globe, including hospital servers and equipment. Through a type of ransomware, the attackers demand payment before releasing targeted machines, and altogether receive more than $140,000. Just one of an increasing number of recent cyberattacks, WannaCry reveals the long way that cybersecurity still has to go.

February 5, 2018: The **American Bar Association** (ABA) officially **accredits privacy as a new legal specialty**, a rare designation. Through the associated certification program, a U.S. attorney who passes the CIPP/US exam, has "substantial involvement" in the field, and satisfies five other requirements, can become a "privacy law specialist."

March 23, 2018: The **CLOUD Act is signed into law**. Among other provisions, the act authorizes and streamlines access of individuals' electronic communications by law enforcement, even if that data is stored abroad. Moreover, it requires American communications-service providers to cooperate with governments partnered with the United States via executive agreements. The act receives strong support from tech companies for its clarification of legal data obligations, but civil liberties organizations worry that it dangerously empowers the executive branch without providing sufficient review.

April 26, 2018: For the first time in several years, the bipartisan **FTC** is **restored to full capacity**, after President Trump's nominees are approved by Congress. This comes in the wake of increased attention to privacy policies, particularly with Cambridge Analytica's data mining of Facebook prior to the 2016 US presidential election. With a renewed FTC, America may be on track to review and implement new privacy guidelines.

Tomorrow and tomorrow…

Chapter 2: Location Tracking and Locational Privacy

Key Questions:

- What can be learned from locational data? Who can benefit from this knowledge? How dangerous is this knowledge to the subject?
- Is location information of a certain resolution personal information? How high must this resolution be to be valuable?
- Who, if anyone, ought to know a citizen's location at any moment?
- Who has access to location information today, and how is it being used?
- How plausible is anonymization as a security solution?
- For locational privacy, how can we resolve the gaps between what is and what ought to be?

> Consider: A new smartphone game, Avian, has seized the population. Via the app, users can search their geographical area for avian "friends." Using population mobility trends, the game creators have scattered virtual birds throughout the world in both popular and less frequented locations. Some of these spots are in the tamed wilderness. Others are strategically placed near retail stores of partner companies to the app makers. By visiting these locations and then "collecting" as many virtual birds as possible, players acquire rating points and gold, currency that can subsequently be exchanged for in-game prizes and improvements to game avatars. In addition to the fictional birds, the game offers an extensive encyclopedia with information about local avian fauna that users can read at their leisure.

Upon opening the app, Avian prompts users to enable their smartphone GPS. In doing so, users can view the profiles of others logged in nearby, although they can alternatively opt to play incognito. Moreover, the app sells anonymized versions of this user location data to interested advertisers. Based on the data, partner stores market services and products to game players.

Is it right that the app and advertisers know location information? What resolution of information is acceptable for transmission?

—Inspired by Pokémon Go

An old proverb tells us not to judge another until we've walked a mile in his or her footsteps. As this ancient wisdom suggests, those steps add up to an intimate portrait, far more informative than a single incident. Broadly, our environment shapes us, sculpting our ideologies and behaviors through life. It also affects our immediate and long-term possibilities, informing the available resources and opportunities.

Location information today is more widely broadcast and collected than ever before. Smartphones are one common mode of collection via location-based services (LBS). As their name implies, these services use location information to perform various tasks—some essential, some convenient, some surreptitious. Telecom companies, for instance, rely on location data to route cell phone calls. Navigational applications such as Google Maps and campus-based food apps (e.g., Foodler) use users' locations to find nearby eateries and directions there. In order to provide accurate guidance in real time, such services identify a user's location via direct access of device coordinates or manual entry. Other apps and services provide automatic weather warnings, identify nearby events of possible interest, advertise for retailers through geo-fencing, and enable emergency services.

In the United States, LBS have rapidly propagated alongside smartphone penetration. From 2010 to 2016, US cell phone use has increased from 20.2 to 63.9 percent.[6] In concert, the dominion of LBS has stretched from merely four percent of adult smartphone users in the United States to ninety percent of all US users.[7] Across the globe, LBS are substantially increasing in other countries as well.[8]

---

[6] See Statista, "Number of Smartphone Users in the United States" and Worldometers, "US Population (Live)."

[7] Zichur, "4% of Online Americans Use Location-Based Services" and eMarketer, "Most Smartphone Owners Use Location-Based Services."

[8] In France, for instance, more than 60 percent of the citizens had smartphones in 2016, up from only 10 percent five years prior, as per Statista, "Number of

Prominently gathered from smartphone navigation apps, location information is also collected and employed on other platforms. Network usage from any device—laptop, smart watch, or phone—can reveal location, such as through the connected Wi-Fi port or the port signature of a dial-up connection. Geographical data collected in this way is often used to personalize web advertisements. Users can also voluntarily reveal their locations. Twitter, for instance, has a feature so that users can stamp their location into posts, while the summer 2017 upgrade of Facebook-owned messaging service WhatsApp included a location-broadcast feature so that users could inform friends and the public of their whereabouts.[9] Not all methods of location identification are equally precise, and to properly evaluate LBS benefits and drawbacks, we should distinguish key types.

Tracing Location Identification Techniques

Some methods of tracing location data are more accurate, some more precise than others. The simplest involves voluntary user-provided information. An automated and more reliable method uses cell-tower-based calculations. Using at least two cell towers, and based on the radio waves of the device, location can be triangulated to within thousands or hundreds of meters. This is how telecom companies provide users with service and bill customers appropriately. Another technique employs Global Positioning System (GPS) technology through a GPS chip embedded in the device that relies on Earth-orbiting satellite signals. Although providing greater precision, within tens of meters, this method is also slower, currently taking up to thirty or more seconds to bounce between satellites and the device. Moreover, this method tends to be less

Smartphone Users in France."Similar trends can be found across Europe with even sharper trends in Africa and Asia.

[9] After testing the location-broadcasting upgrade, WhatsApp released the latest version in the summer of 2017: Perez, "WhatsApp Tests Real-Time Location Sharing."

precise when a device is indoors and beneath stone or wood roofing and floors, accurate only to within hundreds of meters. Some researchers continue to work on IPS (Indoor Positioning System) as a fix for more precise indoor coordinates, but such projects have had limited success and have yet to be integrated beyond local networks.[10] Each of these methods can also be combined, such that information from one method can corroborate or modify the other, and combination tends to yield the most precise results.[11]

Beyond GPS and cell towers, Internet connections are encoded with location as well, situating the "smart" devices that rely on them. Part of the Internet of Things discussed further in Chapter 7, smart devices including phones, watches, spectacles, vehicles, and laptops now come equipped with an impressive amount of computing power. Increasingly, such devices also incorporate GPS chips, microchip RFID tags, as well as adapters for Wi-Fi, Bluetooth, or other network connectivity. To connect to an online network, a device must pass through an access point, and default connections are typically chosen based on port proximity to the device (as this positively correlates with connection strength in general). While new ports arise every day, the most frequented ones tend to be stable, and extensive databases of such Wi-Fi access points have been compiled by Google and others.[12] Similar

---

[10] Radio frequency identification (RFID) tags are one popular use of IPS, particularly in retail. These are microchips thinner than a penny, viable within hundreds of meters. CNRFID, "Introduction to the RFID," provides a somewhat lengthier introduction. Jablonski, "Getting a Fix on Indoor GPS" and Streich, "Technologies for Server-Based Indoor Positioning" provide more on other recent IPS efforts.

[11] Lawson, "Ten Ways Your Smartphone Knows Where You Are" further describes several of the methods mentioned here. It also includes a few others such as ultrasonic near-field communication (NFC) increasingly being used for payments and incorporated into smartphones. New methods continue to be developed, while precision of established methods is also improving.

[12] Though from 2013, the discussion in Horowitz, "Google knows nearly every Wi-Fi password in the world" remains relevant. Also see the "Further Exploration"

to how cell carriers identify location based on triangulation, via a Wi-Fi database lookup, an Internet service provider can then identify a user's location with reasonably high resolution.

Smartphone apps such as Avian that employ LBS typically rely on more precise techniques, commandeering the phone's ingrained chips or accessing network information. This allows for more efficient navigational results, but also means that users' locations are exposed to app developers and data brokers, the subject of Chapter 3. In another way as well, consumers' locations may be exposed: in recent years, technology giants have moved away from applications that readily display user locations (e.g., Google Latitude, Windows 8, and Nearby Friends).[13] Yet, overt display is making a comeback through platforms such as WhatsApp and Twitter. As with most services, there are positive aspects of location sharing, in addition to concerning ones. For the Avian app that began this chapter, both sets are undoubtedly in play.

A Case For Restricted Collection of Location Data

In a November 2010 report, the California branch of the ACLU asserted that people "have the right to expect that new technologies will improve their lives without invading their privacy."[14] Yet, given the rate

---

section of this chapter for how developers and regular consumers can employ the accumulated information.

[13] Google discontinued Latitude in 2013, which allowed users to share their movements over time with friends: Google, "Latitude has been retired." The service has been incorporated into subsequent Google products but with less functionality. Windows 8 from 2012 allowed apps to access a user's name, location, and profile picture by default: Pinola, "Turn off Windows 8's default sharing." In 2016, Facebook shut down Nearby Friends' feature that allowed users, for a time, to share their location with high-precision: Constine, "Facebook kills off exact location sharing."

[14] ACLU Northern California, "Location-Based Services."

at which technologies are developing, as well as the state of current policies, this expectation is not being met. Instead, the business environment is one in which companies such as Facebook can hide the full and dangerous implications of collected data. Illustrating some of these broader issues, the Avian app, like numerous others, warrants grave concern. Ultimately, given the enormous amount of information compacted within the data, its propensity for abuse, and the impossibility of anonymization, location data collection and use must be curtailed.

What Location Reveals Today

Whether precise GPS coordinates, Military Grid Reference System (MGRS) coordinates used by NATO-member militaries, or residence ZIP codes, all location information is personally identifiable information (PII). The category of PII includes specific identifiers (e.g., an individual's name or social security number) as well as indirect identifiers (e.g., an individual's job title or location coordinates). For privacy advocates, PII is a form of personal data, the fundamental element which warrants protection.

Nonetheless, not all PII is equal, and locational data holds special significance because of its implications. Certainly, for an ordinary citizen, public exposure of some sort is inevitable. In fact, perfect anonymity is implausible and assuredly undesirable for the sake of all parties including other citizens, the sociopolitical body politic, and the citizen herself. Extrapolating from this difficulty and pointing to a history of public interactions and open doors, some critics of locational privacy might brush off altogether the immense significance of locational data collection—they would do so unwisely.

Skeptics of location's worth might point to the townships of Puritan New England in which America's early political foundations were forged. Here, intimate communities nourished their own ecosystems, individuals earning their own livelihoods yet closely engaging with one another. Sooner or later, Goodie Proctor's infidelities became common knowledge for local inhabitants. Financial hardships of the Bradford family would be noted by town elders. When a merchant embarked on a business trip, the whole town might join together to wish him well and send him off, aware that his home and property would rest abandoned for the coming weeks. Functionally, such knowledge would not merely serve prurient interests; rather, when better informed, the townspeople, lacking a centralized government and comprehensive set of laws, could coordinate to impart wisdom and aid by committee—or condemn and exile by majority assent.[15]

However, the past is far-removed. As with other arguments in privacy, past developments do not easily translate to modern times, and there are fundamental differences between eras. For one, society today moves at a much faster pace, owing to shiny technologies and enhanced knowledge. Screwdrivers, for instance, now come in many different shapes and sizes—some are even electric; new chemicals act more quickly to incapacitate a target, and fatal doses may be smaller in quantity than the head of a pin.[16] Moreover, the volume of interactions today is substantially greater given that the population of Earth is 7.5 billion and growing, compared with the five hundred million of 1650, when the Puritans arrived on New England's rocky shores. Relatedly, with the

---

[15] The impetus for public knowledge was undoubtedly also influenced by the strong religious views of the Puritans, for whom personal and group piety were fundamental. The community's views naturally evolved with the economy, and as more diverse individuals came to share their cities. Merchants in fact were not well-regarded, at least initially. For more on the societal changes, see Bailyn, *The New England Merchants in the Seventeenth Century*.

One of the more famous Puritan exiles, Roger Williams is generally considered the founder of Rhode Island where he fled after his radical declarations against the Church of England drew the ire of Massachusetts Bay colonists. An analysis of his impact as well as the evolution of American ideals during the course of his life, can be found in Barry, *Roger Williams and the Creation of the American Soul*.

[16] For a quick sampling of non-potable screwdrivers see, Taylor, "5 Types of Screwdrivers Every DIYer Should Get to Know." For more on ricin, see Miller, M, "What the Heck Is Ricin?"

Internet at our fingerprints, society is connected beyond physical boundaries as never before. The significance of any data today hence cannot be equated with data from the slower, sparser, and less leaky past.

Furthermore, location data collected en masse for an individual can be extremely telling. Single data points reveal where an individual is or is not, and can be used to identify when, where, and how to potentially strike a target. Collected across time, locational data also reveals an individual's mobility patterns. When overlaid upon a global or local map—as can readily be done through Google Maps or other navigational software—the data can identify specific spots frequented by that individual, such as favorite clothing stores, parks, or restaurants. Mobility data has also been closely correlated with occupation and financial circumstances,[17] and hence by collecting locational data, these mobility trends can be extrapolated to construct behavioral profiles with gross accuracy. These profiles can—and are—subsequently used to construct hidden psychological profiles and tailored business models, as further discussed in Chapter 3.

In the other direction, mobility data can alternatively pick out an individual from within a larger group. For instance, in one recent study within a European city, four spatiotemporal data points with the resolution of a cell carrier's antenna were sufficient to uniquely identify 95 percent of individuals tested. As this recent research through the MIT Media Lab and affiliates suggests, human movements tend to be unique.[18] Hence, even if supposedly anonymized, location data is not necessarily secure. Whether sold to shady data brokers or internally used by data collectors, this locational data hence informs consumer profiles that howl somewhere in the wilderness, waiting to be rounded up and wielded for predation.

---

[17] See, for instance, Singh, Bozkaya, and Pentland, "Money Walks."

[18] Montjoye et al., "Unique in the Crowd."

One possible vector for abuse stems from the government. Expanding data collection still further to multiple individuals, locational data can serve as a basis for constructing robust social networks, valuable for advertisers seeking to identify key influencers as well as for security operatives. As exposed in the 2013 Snowden leaks, the NSA's once-secret program CO-TRAVELER, makes pivotal use of cell phone locational coordinates. The NSA reportedly notes repeated intersections between a suspect and other individuals, in order to begin constructing an operative network.[19] This collected data is, of course, not public information—public exposure would defeat the program's purpose—but the data is compiled into databases that appear to have no protocol for data destruction.[20]

To some, the collection of such data for security purposes may be a fair trade, but a danger always exists that those in power will further personal vendettas against citizens, an issue discussed more thoroughly in Chapter 6. Nonetheless, other sorts of collectors are indubitably more suspect.

These suspect collectors traffic with shadowy data brokers. For instance, the corporations behind certain apps, including last year's wildly popular Pokémon Go,[21] collect location data on users, which they often sell to data brokers who compile portfolios on individuals. These data sets can then be sold to interested buyers, with little to no impunity under current US law. And cases of obvious misconduct are prolific. In the recent past, data brokers have compiled and sold obviously dangerous lists, including "Oldies but Goodies," which featured the names and contact information for five hundred thousand frequent

---

[19] *Washington Post,* "How the NSA is Tracking People Right Now."

[20] Schneier, *Data and Goliath,* 66.

[21] For instance, see Olivarez-Giles, "'Pokémon Go' Creator Closes Privacy Hole."

gamblers over the age of fifty-five.[22] The Avian app, despite educational benefits to increase ornithology appreciation and enhance climate preservation, appears to be cut from the same cloth as these dangerous breeds. Furthermore, other companies (e.g., Google or Facebook), also discussed at greater length in Chapter 3, internally stockpile consumer information that includes mobility patterns. As such, they are vested with dangerous power.

The current privacy ecosystem is not one friendly to individuals seeking protection. With Internet service providers now able to sell user data with impunity, and other protections including net neutrality crumbling before our eyes, the propensity for abuse is high.[23] The market is being wildly deregulated, with actual cases of abuse proliferating, and the potential for abuse increasing even faster. This dangerous climate must change if democracy, dignity, and justice are to prevail.

Locational Data Is Uniquely Dangerous

Even if locational data collection were permitted in some capacity, a citizen's home—whether a mansion, single-room apartment, or cardboard box on the street—ought to be sacred by default. This is, after all, one of the chief protections for which the American colonists fought for independence. The Third Amendment of the Constitution, preventing soldiers from quartering in private homes during peace time and the Fourth Amendment protecting "persons, houses, papers, and effects" from "unreasonable searches and seizures" directly establish the privacy of the home.[24]

---

[22] Duhigg, "Bilking the Elderly, With a Corporate Assist."

[23] S.J. Res. 34, passed in late March, effectively granted ISPs free rein regarding use of consumer data. See this chapter's "Further Exploration" section for more on net neutrality

[24] *Riley v. California* (2014) is a recent landmark case by which the Supreme Court unanimously ruled that the digital contents of a cell phone are a protected "effect." Nevertheless, the meaning of an "unreasonable search," prohibited by the Fourth Amendment, remains contentious, particularly as the War on Terror has enhanced

Yet, the current environment offers ample opportunities for abuse. Through irresponsible location broadcast apps, home-location data can now be loudly projected. It can be accidentally revealed when a user forgets to turn off the GPS location projection on her smartphone. Alternatively, it can be sneakily siphoned without the user's knowledge, by an entertaining smartphone game, notification of the location theft buried beneath legalese.

In turn, the 30 percent of users who reveal home addresses on posts[25] also open themselves up to identity theft, stalkers, and thieves. In one 2015 case in California, a stalker used Facebook and Instagram photos to deduce whether occupants were at home; if they were not, he

---

fear and interconnected technologies have permeated. Moreover, protection of basic home-address data is legally contentious in the United States, in part given that addresses are a matter of public record in many towns. Towns and states typically employ some protection of address data, such as requiring a physical presence at the town hall to grant records access. However, subsequent broadcast of these records through news stories or, more recently, Twitter messages, is not obviously prohibited.

Moreover another important condition for privacy of public records makes disclosure easier: to be liable for a disclosure suit, the published facts must be "not newsworthy," itself an ambiguous phrase. Ironically, once disclosed, even if not in a legally justified manner, by virtue of public interest in the suit itself, disclosed private information can become newsworthy. In turn, courts have ruled that home addresses of crime suspects are in fact newsworthy and hence can be published with impunity as discussed in Digital Medial Law Project, "Publication of Private Facts." The harmful impact of such publication, however, cannot be understated. One recent case involving the parents of George Zimmerman is discussed is discussed in Duke, "George Zimmerman's Parents Sue Roseanne Barr over Tweets." After Zimmerman shot Trayvon Martin in 2012, inciting furor over racial discrimination, Roseanne Barr tweeted what she thought was George's address to her 110,000 followers. In fact, it was the address of his parents, where he had not lived in six years. The tweet forced the residents to abandon their home amidst repeated threats of retributive violence. Despite the turmoil faced by the parents, their lawsuit against the disclosure was quickly dismissed.

[25] McAfee, "Oversharing on Social Media Common Amongst 50+."

would enter their homes and steal their underwear.[26] I another incident from February 2017, English professional footballer John Terry had his mansion burgled after posting an Instagram photo showing that he and his family were on holiday. Though the thieves were caught, none of the items (worth 400,000 pounds, excluding any sentimental value) were recovered.[27] Members of the paparazzi also stalk celebrities, using posts that reveal location data on social media. Moreover, breach of collected private data—including locational information—has led to a dramatic increase in identity theft, as documented by the public-interest research center EPIC.[28]

Powerful technologies employed by the military and police are exceedingly dangerous if they fall into the clutches of the unauthorized, as they inevitably seem to do. One relatively new technology called the stingray is officially—and still controversially—used only by law enforcement; it would certainly be a grave danger in criminal hands.[29] Simulating a cell tower, the stingray diverts the attention of any targeted mobile device in order to pinpoint location. At the same time, the stingray creates an illusion of network service, effectively isolating the target from others. Similarly, a hack into a database of Wi-Fi ports or a telecom company would yield fruitful location results in real-time.

Certainly, if we wish to continue using the modern technologies upon which we rely—cell and wireless services, among others—LBS

---

[26] Serna, "Man Allegedly Used Facebook, Instagram to Find Women, Steal Their Underwear."

[27] Terry's post was captioned simply "Great few days away Skiing with the family" with a picture of him and his wife on the French Alps (Telegraph Reporters, "Designer goods worth £400,000 stolen from John Terry's mansion").

[28] See the "Further Exploration" section of this chapter for links to recent EPIC news updates.

[29] Farivar, "How Florida Cops Went Door to Door with Fake Cell Device to Find One Man."

cannot be abandoned altogether. However, the significance of location information and current trends of abuse highlight that far more must be done to defend consumers. More vigilance is needed in monitoring LBS, and more protections need to be established to reduce its risks.

SPOTLIGHT: CAN CELL PHONES BE SEARCHED WITHOUT A WARRANT?

In *Riley v. California* (2014), the US Supreme Court unanimously established that the warrantless search and seizure of a cell phone's contents during arrest is unconstitutional, a sharp contrast to the longstanding precedent of *Chimel v. California* (1969). The court's verdict signaled that in the eyes of the law, cell phones had to be treated as a special sort of device with graver privacy protections attached. An ongoing case hailing from Cincinnati (*Carpenter v. U.S.*) is currently on the Supreme Court's fall 2017 schedule and will be one to watch. In *Carpenter*, the defense for a man convicted in a series of armed robberies has claimed that warrantless search of a cell phone's location data is an "unreasonable search and seizure" (Hurley, "US Supreme Court to Settle Major Cellphone Privacy Case"). However it is resolved, this case will develop an important precedent for the privacy of smartphone and possibly Thing device data.

The Risk of Re-Identification

Data anonymization is sometimes offered as an adequate defense for data collection. The argument runs that if personal information is anonymized then gross trends can be deciphered to benefit sociologists and marketers without endangering individuals. However, re-identification, even of anonymized or pseudonymized data, has been demonstrated to be quite easy. This is significant specifically for location data and the data to be uncovered by Avian, but the implications stretch well beyond into all personal data.

The MIT study mentioned above illustrated that very little

locational information can reveal a lot. The idea stretches well beyond. A landmark study by data-science pioneer Latanya Sweeney in the 2000s suggested that three simple data points—date of birth, ZIP code, and gender—could uniquely identify someone in the United States by name eighty-seven percent of the time.[30] Her first documented foray into re-identification was even more discerning, when she was able to accurately identify the health records of Massachusetts Governor Weld. At the time, Massachusetts released hospital data as a means of public disclosure and their anonymized data was understood to be sufficiently protected. Combining Weld's data with voter-roll data purchased for just twenty dollars from public databases, Sweeney undermined the legal wisdom: she was easily able to identify Weld's diagnoses, prescriptions, and other details.

In the immediate aftermath, others pushed back. Epidemiologist Daniel Barth-Jones, for instance, has argued that Weld's re-identification was only possible because the target was a public figure with highly publicized hospitalization, rather than fundamentally informative data.[31] However, public information and databases have only expanded since the Internet boom of the early noughties. With the advent of social media and with increased engagement in online games and interconnected phone apps, whether with pseudonymous or veritable usernames, personal data of any consumer flows wildly. If Governor Weld's information was only identifiable because of his prominent stature, Sweeney's subsequent 2000 study in which she identified 87 percent of her sample set indicated it was no fluke. More recent data analysis efforts such as those by Law Professor Paul Ohm in 2010, MIT Media Lab researchers in 2011, and *New York Times* reporters in 2006

[30] Sweeney, Simple Demographics Often Identify People Uniquely 2000.

[31] Barth-Jones, "The 'Re-identification' of Governor William Weld's Medical Information."

who were data analysis amateurs[32] reveal the continued relevance of the re-identification problem.

Another tactic used to undermine the prospect of re-identification has been to emphasize anonymized data's societal value. In 2011, then-law student Jane Yakowitz highlighted how data could enlighten by ripping out, for instance, roots of support for discriminatory beliefs. Moreover, turning re-identification on its head, Yakowitz plausibly championed the concept of information justice whereby access to information is a fundamental right for every citizen. After all, data can offer immensely valuable insights that improve health and professional opportunities, and it is not fair for only some to have access. For instance, data can identify the effects of smoking on fetuses, identify the nature of crime in particular neighborhoods, determine how police resources are apportioned, and identify the prevalence of particular fraud schemes. All of these have broad interest. Fundamentally, Yakowitz also noted that "collectively, our data reveals more than any of us could know on our own."[33] For all these reasons, she argued that a better and more just future would necessitate the open sharing of data sets. The alternative would be privatized information troves. In turn, oversight would inescapably be limited, enabling both honest mistakes and intentional release of selective, self-serving data.

Yakowitz's incisive arguments highlight the great value of data to all members of society, an idea to be revisited in chapters 7 and 9. However, when it comes to the data dossiers compiled today by data brokers and the locational data collected today by governments and corporations, these follow the privatized model that Yakowitz rightfully criticizes. Governments may utilize the data for legitimate national

---

[32] Respectively, see Ohm, "Broken Promises of Privacy: Responding to the Surprising Failure of Anonymization," Montjoye et al., "Unique in the Crowd," and Schwartz and Solove "The PII Problem," 1841.

[33] Yakowitz, "Tragedy of the Data Commons," 13.

security purposes, but if corporations simply produce profits for themselves, this does not benefit society.

More significantly, the solution would not be to simply broadcast all personal data. Yakowitz and others argue that re-identification techniques are neither as easy nor as successful as vocal critics have suggested. Certainly, not all useful data necessarily violates privacy, and not every variable in public data sets is an indirect identifier. Nevertheless, as the examples from Dr. Sweeney's, Dr. Ohm's, the MIT Media Lab's, and the *New York Times* reporters' research indicate, it truly is the case that a few pieces of data are sufficient to reveal a lot.

With the expansion of networks, personal data shared on social media today can accelerate deep into the web, accessible by acquaintances of acquaintances and perhaps even the public-at-large. Stalkers, abusers, and the unhealthily curious garner access as never before. New revelations keep emerging as well. A February 2017 study at Japan's National Institute of Informatics suggests that pictures are in fact worth a great many words: high-resolution photographs—the sort that most smartphones currently take—featuring an individual's palms were used to accurately reproduce fingerprints.[34] Meanwhile, the fall 2017 Apple iPhone X relies on facial recognition of an authorized user to unlock the screen. The technology underlying this, FaceID,[35] is part of a broader push towards compiling and using all sorts of personal data.

The more personal data proliferates, the easier it becomes to link up disseminated elements and re-identify individuals. Anonymization—and even the lower bar of identity cloaking—will become an increasingly impossible feat. Nevertheless, phone apps and other sources simply continue collecting and selling data today. Perhaps, even after noting all of the risks, many users will still be willing to share their geographical coordinates—after all, there is no such thing as a free lunch, and the

[34] McGoogan and Demetriou, "Peace Sign Selfies Could Let Hackers Copy Your Fingerprints."

[35] Greenberg, "How Secure is the IPhone X's FaceID?"

societal benefits described by Yakowitz remain valuable. Nevertheless, it should be a user's active decision to bequeath such information; leakage should not be the default.

In some cases, users' data leakage may be inadvertent on the part of app developers, the product of poor security measures. Even otherwise reputable companies have fallen prey to this. For instance, in 2014, the highly popular Starbucks app as well as the popular Coupons App were both discovered to store user passwords in clear text, and the latter was found to actually transmit location coordinates across the Internet with each use.[36] Stumbling across such information, a stalker could easily use it to perform unwanted purchases and commit identity theft. Further difficulties arising from reliance on interconnectivity and cloud computing will be discussed in Chapters 7 and 8. In any case, it is clear that data leakage, of locational coordinates and more, are quite significant.

LBS and other services reliant on data are unlikely to vanish. Hence, the practical solutions—beyond renouncing technology and moving off the grid—rely on improving best practices and instituting legislation to improve security measures as well as consumer choice. With regard to best practices, whereas the GDPR passed by the EU establishes clear requirements, Congress has yet to develop comprehensive legislation. Some fair information practices exist in specialized fields such as medicine (via HIPAA, as will be explored in Chapter 9) and banking (the FCRA and GLBA), but on the whole, few regulations are in place and few limits exist on collection for the vast majority of interactions. The impetus for change is nigh. In 2012, the Federal Trade Commission (FTC), which has historically overseen domestic privacy law, released a comprehensive report on "Protecting Consumer Privacy in an Era of Rapid Change." Among other

[36] Schuman, "Starbucks Caught Storing Mobile Passwords in Clear Text" discusses the Starbucks leakage, while Eddy, "Android Coupons App Leaks Your Personal Information to Everyone" expands on the Coupons App.

recommendations, the FTC called for Congress "to provide clear standards and appropriate incentives to ensure basic privacy protections across all industry sectors."[37] Importantly, the FTC asserted that this baseline legislation could preserve corporate flexibility and continue promoting innovation. Nevertheless, a passive Congress has yet to pass any such bill.

We have powerful incentives to share data and profit as a society. Anonymization and pseudonymization are some means of doing this, but are not entirely secure. It seems likely that cybersecurity will remain a perpetual battle between aggressors and defenders, and as such, anonymization techniques that appear secure at one point in time—limiting information to merely zip code, birth date, and gender, for instance—will have to be consistently re-evaluated. At the very least, the ease of re-identification means that current protections of personal data in the United States that prohibit use of direct identifiers such as name, home address, or phone number and form the basis of many regulations are incomplete and insufficient. More effective risk-mitigation techniques can still be implemented, including, as Yakowitz notes, layered security measures to make re-identification so time-consuming a process as to be impractical.

Fundamentally, the current environment of location privacy is one ripe for abuse. Consumers have some control over the dissemination of their information, and to the extent we can, we must be careful about what permissions we allow. Particular tools and techniques—such as VPN networks[38] and others highlighted in Chapter 8—can further increase protection by reducing data leakage. However, through surreptitious and sneaky data collection permitted by current legislation and standards of practice, advertisers, corporations, and their clients will still have access to highly revealing information.

---

[37] Federal Trade Commission, "Protecting Consumer Privacy in an Era of Rapid Change," 12.

[38] See this chapter's "Further Exploration" section for more on these networks.

Citizen choice is essential to privacy, and there are some benefits of LBS that make blanket prohibition of location data collection unacceptable. There is much middle ground, however, between prohibition on one side, and the open season of the present. With our locational data usurped into the shadows and the chilly winds whipping our coordinates away from us, now is the winter of discontent. Through the light of transparency, through sensible regulations as fertilizer for consumer interactions, let us make it a glorious summer, before the opportunity is past.

A Case For Wide Collection of Location Data

It certainly is the case that valuable location data raises a privacy risk. However, the critic above does not adequately acknowledge the myriad benefits of LBS. Among other benefits, bequeathing location information provides users with notification of nearby services for work and leisure, increased possible contact with companions, and essential emergency services. While the behavior of certain data brokers with lists such as "Oldies but Goodies" is unacceptable, this is not the norm for LBS, nor does rectifying abuse require drastic overhaul. On the whole, to truly respect the choice inherent to privacy, to satisfy the necessity of corporate profit, and to enable superior emergency services, locational data collection today requires only minor tweaks to best benefit society.

The Necessity of Choice

As the critic notes above, citizen choice is essential, and such choice precludes blanket prohibitions or permissions. Choice in privacy is not unlike the choice to engage in the social contract that forms the basis of modern political states. Many countries grant either *jus soli* or *jus sanguinis*. In these cases, a person born on a nation's soil, or born of a citizen, respectively, becomes a citizen. From birth, that citizen is entitled to various rights. While the UN Declaration of Human Rights boldly establishes rights for all *Homo sapiens*, it is typically on the national level that individuals have enforceable obligations. Failure to engage in these can be punished; for instance, driving on the wrong side of the road or driving while intoxicated can result in fines and jail time, even if no other person is directly harmed. In certain locales, these obligations can extend further to personal property. For instance, one might be prohibited from constructing roads on one's property without town approval due to concern for fragile wildlife. Hence, some restrictions infringe upon individual rights, but they serve to protect society's interests in general. Ultimately, citizens have some choice in abiding by society's regulations as well: they can challenge a regulation in court, and can even take residence elsewhere to leave the contract altogether.

Certain restrictions on autonomy are hence necessary and acceptable. However, any sort of broad prohibition to locational data collection would not clearly benefit society as a whole. Unlike road navigation and land development, the risks of data theft or abuse are chiefly individual; shared location data does not obviously endanger others. Moreover, individual views on how much to share have been demonstrated to differ widely.[39] Hence, to accommodate those concerned about data sharing, there should be an option such that individuals passionate enough to obsessively protect their personal

---

[39] Olenski, "For Consumers, Data is a Matter of Trust" analyzes the results of an October 2015 Columbia Business School study which found that for eight thousand consumers from five countries (Canada, France, India, the UK, and the US) 75% were generally willing to share personal data with companies in exchange for a valued service. Shared data included address, mobile phone number, name, and date of birth. Nevertheless, willingness varied by industry considered—companies in finance, for example, were greatly trusted—as well as by age group polled. A more recent study, GfK, "Willingness to share personal data in exchange for benefits or rewards," was conducted by one of the world's top five market research institutes in the world. Compiled in the summer of 2016, the study polled twenty-two thousand consumers in seventeen countries with a graded scale of willingness to exchange personal data for corporate rewards. Here as well, substantial deviations were apparent. Globally, 27% were absolutely willing to exchange personal data for rewards, but this included only 8% of Japanese consumers, in contrast to 25% of American responders, and 38% of Chinese interviewees.

information should be able to do so. On the other hand, those willing to exchange personal data for services should also be enabled to continue doing so.

Moreover, Internet service is not self-sustaining; search engines, the means of navigating cyberspace, are private companies. Financially, while we may have come to depend on services such as web access and phone apps these and other often-free services cannot self-subsist. Simply in order to survive, the companies that provide these services require something in return, and given the immense value these companies provide, profits well above the margin seem well-deserved. Such profits will feed into still other benefits, including innovation that leads to greater consumer choice.

ISP Autonomy and Innovation's Rise

Particularly for telecom companies providing wireless services, a rather abstract yet crucial debate—that of net neutrality—lies at the core. Internet service providers (ISPs) supply consumers with Internet connections. Within the United States, there are thousands such providers offering Wi-Fi, dial-up, and other cable connections. Most service local areas, but a few giants including AT&T, Comcast, and Verizon have broad reach. The services these companies provide have come to be essential, depended upon by schoolchildren, independent researchers, businesses, and more. In fact, a wireless network is arguably even richer in value than basic commodities like water and oil since it provides access to the Internet's bounty. And through the web, like new fishermen, users can learn to procure water and oil for life, rather than merely having a day's or week's supply. Given the power of the web, it seems right to note as Yakowitz and other data devotees have done that access to the web and its data should be considered a basic human right.

Naturally, then, the question becomes how can access best be promoted to maximize opportunities for all? One faulty but popular answer focuses keenly on present problems. It points to how other fundamental industries came to be managed by the federal government.

In 1887, the railroads became the first industry subject to broad regulation. Rapidly growing, and the chief means of transport for goods and people at the turn of the nineteenth century, railroads were also dominated by monopolies. Abusing their power, these companies instituted polices that seemed arbitrary and moreover appeared to discriminate unjustly against small businesses and other possible competitors. Through the Interstate Commerce Act, Congress sought to regulate such "common carriers" and increase just opportunities.[40] Over the next forty years, the notion of a "common carrier" evolved, coming to include oil and gas pipelines as well. Then, in the wake of the Great Depression, with World War II looming on the horizon, telecom companies began to grow in the United States. Through the Communications Act of 1934, President FDR oversaw the installation of telephone poles and radio towers across the country, in part to usher in new jobs, in part to bolster national security, and in part to champion innovation. Among other provisions, the act created a regulatory body called the Federal Communications Commission (FCC) that would regulate interstate communications to ensure just service. At that time, telecom companies were also categorized as "common carriers" and placed, through that designation, under FCC jurisdiction.

When the Internet boomed in the 1990s, it was supplied through specialized network lines. These lines ran from public spaces to individuals' homes to provide early Internet connections. Largely in an unregulated state, ISPs happily expanded their reach and services offered to citizens over the next decade. Since networks required wired connections, telecom companies, already having established lines in the past century, began to stake out larger chunks of the ISP industry. Then in 2015, ISPs were suddenly classified as "common carriers" by the FCC

[40] For more on the Interstate Commerce Act, see United States Senate, "The Interstate Commerce Act is Passed," and "Interstate Commerce Act."

and thus placed under clear government oversight.[41]

With the 2015 FCC designation, the agency hastily promoted "net neutrality." This idea, attributed to Columbia law professor Tim Wu, asserts that ISPs should treat all data on the Internet equally and not restrict access to certain individuals or prioritize particular content in which the ISP might be invested. Certainly a reasonable idea, the vocal furor in its favor nevertheless has been premature. In some basic respects, the state of the web today may mirror the state of railroads in the late 1800s. As there was then, there is a novel and powerful technology at hand, with a few players dominant. Moreover, regulations then, as now, are quite minimal. Undoubtedly, the web is a carrier of information and something on which we rely not only professionally, but also personally, as the pathway to knowledge and action. Yet, the differences are far more salient. For one, whereas railroads are bounded in space and options, Internet pathways are far more diverse. This means that whereas railroad regulations kicked in after the railroad system had essentially been built, we are not nearly at such a state with ISPs.[42] For instance, 5G networks promising to be at least ten times faster than the previous iteration approach rapidly, while 6G and future generations are already being considered.[43] Hence, the limits on innovation bound to result from increased regulation sound and look severe.

Even more significantly, data to support claims of ISP abuse

[41] Some of the early ISPs are described at "10 Early ISPs and What Has Become of Them." For the FCC's 2015 classification, see Fung, "Net neutrality takes effect today. Here's how it affects you."

[42] Of course, some innovators may take umbrage at the idea that public transport has come as far as it can: for instance, see Barbash, "'Tesla-in-a-tunnel,'" about the "Chicago Express Loop" that Elon Musk's Boring Company aspires to develop thirty to sixty feet beneath the city. While there may be room for innovative ideas for public transport, the Internet's digital highways undoubtedly leave even more room for future development.

[43] For more on 5G and 6G see, e.g., Nunez, "What Is 5G and How Will It Make My Life Better" and Koziol, "It's Never Too Early to Think About 6G."

remains quite sparse. Consider for yourself the evidence compiled by FreePress.net, a staunch supporter of net neutrality.[44] The site cites thirteen alleged cases of abuse through 2015. Already a rather small number, several of the cited cases are moreover of dubious merit. For instance, Comcast's 2005 block of BitTorrent and other peer-to-peer networks was arguably justified, given how such networks can be used to propagate illegal materials including copyrighted content.[45] At least a few other cited violations of net neutrality were likely honest mistakes. In any case, clearly no violation occurred on a massive scale, and clearly all violations, unintentional or otherwise, were caught soon after they occurred, then rapidly rectified. This speaks to still another significant difference between time periods—there is far more transparency now than in the past. Through a free press and rapid communications, communication incidentally fostered by improvements to web connectivity, citizen activists are empowered to investigate possible abuses and bring them to the attention of others. This requires some time and effort, but it is also the responsibility of citizens to remain active for their sake and for their peers. There is hardly any evidence to suggest that the regime was in 2015—or now is—in serious need of change. Rather, the strongest arguments remain speculative, about possible abuse by ISPs.[46]

Nevertheless, the FCC pushed forward and aggressively

[44] See Karr, "Net Neutrality Violations: A Brief History." The evidence and reasoning are similar to that culled by the American Civil Liberties Union in defense of Net Neutrality: ACLU, "What Is Net Neutrality."

[45] One study from 2010 claimed that less than one percent of BitTorrent files were legal: Cheng, "Only 0.3% of files on BitTorrent confirmed to be legal." Even if the study's results were exaggerated, Torrent remains known for illegal downloads, such as of films that win Best Picture: Winograd, "12 Years a Slave Sees Surge in Illegal Downloads After Oscar Win."

[46] Net neutrality is undoubtedly a complex topic. See Wu, "Network Neutrality, Broadband Discrimination" for the pivotal paper that began the discussion. Also see this chapter's "Further Exploration" for links to more.

published further ISP regulations in late 2016, at the tail-end of the Obama administration. The rules were entitled "Protecting the Privacy of Customers and Broadband and Other Telecommunications Services," a curious concoction since the domain of privacy regulations in the United States has belonged to the Federal Trade Commission (FTC) since the 1970s.[47] While instituting some potentially reasonable rules, such as adoption of particular data security and breach requirements, the FCC nevertheless overreached with this order, prohibiting ISP service offerings that were contingent on customers surrendering certain privacy rights. As addressed above, such restrictions could preclude just profits for corporations providing an immensely valuable service. Moreover, these significant ecosystem restrictions came from the FCC, an executive agency unelected by the people. This attempt at extended executive power was dangerous not only for its potential for abuse, but moreover for the inevitable bureaucratic morass it would create, stunting development.

Thankfully then, the FCC's "unilateral decision" was shut down by Congress a few months later in a close but successful vote (215-205 in the House and 50-48 in the Senate).[48] With *S.J. Res. 34*, ISP authority

---

[47] See Federal Trade Commission, "Protecting Consumer Privacy." The FTC's reign over privacy may seem unusual, given that the organization was originally created in 1914 to break up trusts. Nevertheless, one of the agency's three bureaus is now the Bureau of Consumer Protection, and the FTC professes a mission to "protect consumers and promote competition." In protecting consumers, the FTC has taken an especially active role in digital privacy since the late 1990s, pursuing companies such as Eli Lilly, Microsoft, and Facebook for insufficient security that compromised consumers' personal data, as well as retroactive changes to corporate privacy policies, which it deemed unfair.

[48] See Federal Trade Commission, "Joint Statement of Acting FTC Chairman Maureen K. Ohlhausen and FCC Chairman Ajit Pai on Protecting Americans' Online Privacy," for the new Chairmen's criticism of the late 2016 FCC rules. In their statement, the two agency heads explicitly indicated their support for privacy moving forward, highlighting a massive improvement possible through repeal of the FCC rules: "We believe that the best way to [protect the online privacy of American consumers] is through a comprehensive and consistent framework." Besides placing a burden on the ISP economy, the proposed FCC rules further

was stripped from the FCC by Congress, ISPs declassified as "common carriers." Critics grew agitated in response, some seeking donations for thousands of dollars, allegedly "to buy the browser history of all Congresspeople who voted to sell our data and privacy without consent."[49] These activists claimed that due to ISP deregulation, the gatekeepers could and would peer into consumers' browsing history. These companies, the argument went, could then abuse the data, and soon take advantage of consumers. An FCC vote in the winter was similarly chastised with apocalyptic fervor, but the Restoring Internet Freedom Order carried through, and the agency undid the damages of the officiously imposed 2015 net neutrality movement, thus opening the industry and stimulating innovation for at least the near-future.

Despite the fact that the Order went into effect on June 11, 2018, America remains standing tall, at least with regard to Internet services.[50] Moreover, while critics tend to paint the situation as a stark one, it is clear that consumer interests remain key to ISP developments moving forward. For one, this was encoded into the Restoring Internet Freedom Order, with "Consumer Protection" and "Transparency" the two primary principles, even before Removal of "Unnecessary Regulations to Promote Broadband Investment."[51] Moreover, it remains in all legitimate parties' interests for consumers to be well-served. For the government, consumer choice will drive innovation, and it will also mean that companies receive just profits according to their worth. This will reduce the load for government to monitor, and moreover contribute to

---

fragmented the possible reach of any privacy regulations, by giving privacy-regulating power to the FCC rather than the FTC.

[49] For instance, see Firozi, "Internet users raise funds to buy lawmakers' browsing histories in protest."

[50] Links to the net neutrality rules and regulations described here can be found in this chapter's "Further Exploration" section, along with further related links.

[51] See Federal Communications Commission, "Restoring Internet Freedom."

a happier populace, in turn more likely to be happy with their elected representatives. For the corporations providing the services, their worth rests on public trust. There is sufficient competition across the United States so an abuse of personal data would tarnish first that company's reputation, then its profits.[52] Hence, the current regime shows promise and minimal regulations should continue to be preserved.

Valuing choice does engender some instability. It certainly is possible that moving forward, regulations will need to be implemented: if the majority of society is concerned about ISP access to personal data, then the power rightfully remains in its hands to change course through pressure on the legislature. However, at the present, there are merely unsubstantiated fears, with massive development still possible for Internet services.

The nature of net service providers suggests that they should follow a straightforward economic model. In such a model, available services are determined by needs and success; if individuals dislike some product because it apparently abuses citizen location or other personal data, then they can boycott it so as to stop its business. Incidentally, such a model, rather than some edict from above to outlaw locational data collection by particular companies, respects the core of privacy. It places power in citizens' hands to allow them to make a choice. Only such a model of choice truly allows individuals to exercise their privacy rights.

LBS are pervasive and, as noted by the critic, essential to the functioning of smartphones, laptops, and many other devices today. Hence, any outright prohibition would be foolhardy. Moreover, many of the involved parties who may finance free apps for public consumption (e.g., Google or Facebook) also require profits to function. Presently, "free" navigational apps or games may not be entirely free as they do gather information in exchange for product use. However, a trade is not

[52] For a visual map of ISP competition across the United States, see this chapter's "Further Exploration."

only fair; it's the only way to maintain the economic order. In exchange for the thirty minutes of pleasure that an individual might derive from a game every day, is collection of mobility data for that time frame truly insidious? Individuals can rightfully differ in their opinions, and if they do, they retain the voting voice by which to enact change. However, for the vast majority, exposure of this data, even when illegally revealed via a breach of the collecting company, is probably insignificant. The threat of snoopers is not illusory but a preventive policy primarily shaped by fear of unknown stalkers succumbs too easily to paranoia.

SPOTLIGHT: WILL MOBILE LOCATION TRACKING BE REGULATED SOON?

In June 2016, the Federal Trade Commission (FTC) entered its first settlement with a mobile advertising network (Sannappa and Cranor, "A Deep Dive into Mobile App Location Privacy Following the InMobi Settlement"). The network was InMobi, self-described as "the world's largest independent mobile advertising company." The FTC pursues regulation of "unfair and deceptive practices," and through consent decrees, has previously settled with tech companies including Microsoft and Facebook for diverging from their advertised privacy policies and failing to have adequate data-security systems. In targeting mobile advertising, the FTC is increasing its scope of influence, even while legislation on mobile privacy is lacking.

The Avian app goes beyond even other services in providing educational value, while offering further benefits such as inspiration for exercise. These benefits come in addition to diversionary entertainment and the joy of game-playing. The collection of locational data in this case, easily seems a fair price to pay.

E911: Enabling Emergency Care

The economics of network usage are all too often forgotten by overzealous privacy enthusiasts, but another great benefit of LBS also mandates its continued development: the lifesaving potential of location identification.

Emergency services are not simply a valuable service but also an American tradition. In the early Puritan townships focused on agrarian livelihoods and straw-filled wooden barns, where heating came exclusively from fire, a stray spark and burned bush were not uncommon. With their focus on farming the land, every male community member was also a part-time firefighter who would use his water buckets to quench dangerous flames. As the years rolled by, firefighting became a more regular occupation with the first organized brigade established in Cincinnati in 1853, and spreading since like a wildfire.[53] Over time, improved policies and technologies such as public fire hydrants, improved hoses, and suits have meant that specialization is important to master proper techniques and best practices.

Similarly, other emergency operators have diversified with improved technologies and knowledge, resulting in specialized doctors and full-time police and sheriffs. With the advent of expedited long-distance communication via telephones, emergency responders are another important and established role. Since 1968 in the United States, 911 has been the standard emergency number connecting callers with 24-7 responders who can then send firefighters, police, or ambulances to a particular location. Whether 999 in the UK and Zimbabwe or 112 in most EU nations, emergency phone service is now widely considered an essential provision of the nation state.

For emergency responders, identifying the problem and identifying the caller's location are the first important steps. Unfortunately, owing to the nature of emergencies, which often involve increasingly dire circumstances and increasing hysteria, many callers do not provide coherent locational information.

---

[53] Sometimes, firefighting erupted into competition, as in a dark comedy: Landers, "In the Early 19th Century, Firefighters Fought Fires ... and Each Other."

In an emergency, more than ever, time is of the essence. In precise terms, according to a recent Columbia University study, "a one minute reduction in emergency medical (EMS) response time equates to a 17 percent decrease in the likelihood of 90-day mortality."[54] The medical standard for measuring effectiveness of treatment or intervention, "90-day mortality" refers to the rate at which patients receiving EMS intervention for catastrophic medical conditions (such as heart attack, stroke, or major physical trauma) are likely to survive ninety days after intervention. Inversely, higher EMS response times are correlated with increased rates of 90-day mortality. Further highlighting the value of locational knowledge, recent estimates suggest that for the twenty-five million people who call for an ambulance every year, improved location information could save over ten thousand lives annually (.04 percent of the US population).[55]

Into this bloody fold, LBS via smartphones offer a tremendous solution. Prior to cell phones, enhanced 911 (E911) allowed emergency operators to locate about ninety-five percent of callers via landlines.[56] They employed a pre-established automatic location identification (ALI) database that included regionally collected information. Using this database, maintained by local telephone companies, operators could map a call from a landline to a physical address. Since the address would be static for landlines, operators could easily direct emergency personnel to the individuals at risk.

However, the telecommunications landscape is changing. Along with the rise of cell phones, there has been a decline in pay phone availability and general landline use. Whereas almost every US household

---

[54] Wilde, et al., "Do Emergency Medical System Response Times Matter for Health Outcomes?"

[55] Key et al., "Can First Responders Be Sent to Selected 9-1-1 Emergency Medical Services Calls."

[56] NENA, "9-1-1 Origin and History."

had a landline in 2000, by 2015, forty-five percent of the United States[57] relied solely on cell phones. This is significant, since mobile phones do not have to sit in a designated area, and so the same sort of ALI database is insufficient.

The possibility for emergency services to improve is not merely theoretical, with several recent incidents in various states involving needless fatalities from miscommunication.[58] Taking note of such accidents, the FCC has ordered an initiative whereby cell phone calls transmit forty percent of US callers' latitude and longitude data by 2017 and eighty percent by 2021. Telephone companies have been ordered to meet this by nearly any means at their disposal, and beginning early 2017, companies are publicly documenting their efforts.[59]

In response to the mandate, a critic of LBS might be concerned on three fronts. First, hackers able to breach telecom databases and view call requests in real time will be privy to highly valuable location data. Secondly, even without a breach, criminals could seize and deploy a technology like the stingray to identify the precise location of a target.[60] Finally, the larger question persists over whether a government ought to have access to precise citizen information, due to the potential for abuse and the risk of the Chilling Effect, as discussed in Chapter 6. Certainly, the intended mechanism of tracking for an emergency service is different from mass surveillance in that it is instigated by the user; nevertheless,

---

[57] For the 2000 landline numbers, see Hecht, "FCC Sets Rules for Copper Phase Out."Associated Press, "FCC Prepares Rules for the Post-Copper Phone World" highlights the growth by 2015.

[58] Kelly and Keefe, "911's Deadly Flaw: Lack of Location Data."

[59] The FTC initiative is further described in Federal Communicactions Commision, "Enhanced 911," and Boyle, "Calling 911 on a Cell? They Won't Know Your Address." The latest documents can be found in this chapter's "Further Exploration" section.

[60] For more on the stingray, see Farivar, "How Florida Cops Went Door to Door with Fake Cell Device to Find One Man."

the mere transmission of location data enables another rich facet for potential surveillance. The question in play is hence whether the potential of saving just one life (or, more accurately, the ten thousand featured in estimates) overrides the dangers of a privacy breach to the aggregate. If there is a moral ledger to be balanced, how should this vulnerability risk be quantified?

With the same caveats that manifest for issues of government surveillance, the answer seems to be that every life counts. Hence, development of the technology ought to continue. Emergency service provided by the state is similar to citizenship in a nation. Legally, it is a right that follows from citizenship (and it may also be applicable for those uninitiated), but more significantly, emergency services follow an opt-out model: the default expectation is that a citizen has access to the emergency aid. One can sever connections and live cozily in a cabin in the woods or in an oasis deep in the desert without a phone line, thereby abrogating aid. Yet, this is not the norm; for society's health, the location identification services used by responders are essential.

On account of sustaining the economic model for the sake of innovation, on account of maximizing consumer choice, and on account of defending the dignity of every life, locational data should continue to be collected and utilized. Certainly, locational data wields great power, but this is a power that can be used for good while abuse is restrained.

In Conclusion

In the US Constitution, numerous rights are explicitly enshrined, while the absence of others "shall not be construed to deny or disparage" them. Many of these rights have qualifications and do not grant the rights-holder immunity. For instance, the right to free speech does not mean one can scream "Fire!" in a movie theater or that the expression of racist ideas will have no consequences, such as disapprobation by one's peers.[61] Similarly, a right to privacy does not mean that the right exists absolutely; in fact, in certain situations, such as health emergencies and perhaps even the tracking of terrorist suspects, it can be right to overrule privacy considerations.

The reality of the present is that technology and computing power make it impossible to be truly anonymous. Re-identification is easier now than ever before, and the technology already and soon-to-be developed will not fade away. Nor should it, given the great benefits it can offer. At the same time, though locational and other personal information is widely collected, not every avenue of collection leads to a rightful cause. Hence, beyond improved security techniques, the biggest factor for any change will be grappling and wrestling with the present realities. Greater vigilance on the part of consumers is required, and if it seems significant enough, so is increased transparency such that companies disclose both what data is collected as well as the usage implications of such collection.

Ultimately, privacy—like any right—involves a choice of the endowed. The endowed should have a choice whether to release private data and whom to release it to. To make an informed choice, knowledge of historical risks as well as the state of things at play are invaluable. And ultimately, our personal choices will translate into the policy choices of our body politic. For Avian, as with any other application, defending privacy begins with caring about that choice.

---

[61] For one particularly eloquent articulation of the social pressures that can drive democratic societies, see de Tocqueville, *Democracy in America*, III.21, pps. 615-7. As de Tocqueville considers there, sometimes fear of disapprobation can stultify new thoughts; alternatively, the fear can also dissuade abuses of power.

Further Exploration

The **Electronic Privacy Research Center** (EPIC) maintains a periodically updated site detailing location-privacy legislation that has been recently proposed: **https://www.epic.org/privacy/location**. On its site, EPIC also highlights new threat vectors of identity theft given modern technologies: **https://epic.org/privacy/idtheft**.

Using its accumulated map information (including Wi-Fi ports) **Google Maps** has launched a platform for **developers'** own projects: **https://developers.google.com/maps/documentation/geolocation/intro**. Other maps of Wi-Fi ports are wholly public. For instance, **WiFi Map** is a crowd-sourced project, also available as a free app, through which users have added over one hundred million free Wi-Fi ports across the globe: **https://www.wifimap.io**. One further project is **WiGLE**, which "aims to show people about wireless in a more-technical capacity than your average static map or newspaper article." More information is at **https://wigle.net/faq**.

The **FCC's 2016 privacy regulations** for Internet service providers (termed "BIAS" in the document) can be viewed at **https://www.federalregister.gov/documents/2016/12/02/2016-28006/protecting-the-privacy-of-customers-of-broadband-and-other-telecommunications-services**. These were repealed in the contentious *S.J. Res. 34*, which was signed into law on April 3, 2017: **https://www.congress.gov/bill/115th-congress/senate-joint-resolution/34/text**.

Intimately linked with the regulations above has been concern about **Net Neutrality**, the repeal of which went into effect June 11, 2018. An intense debate still rages around the subject. A mostly objective overview is available via discussion by law professors Orin Kerr and Gus Herwitz: **https://www.washingtonpost.com/news/volokh-**

conspiracy/wp/2017/11/28/one-law-professors-overview-of-the-confusing-net-neutrality-debate. In partnership with Demand Progress and Free Press Action, **Fight for the Future** has created a tool to view where each member of Congress stands on net neutrality. Although the site is clearly intended to pressure representatives to officially vote for a net neutrality bill, the tool also provides a valuable visual, regardless of your perspective: **https://www.battleforthenet.com/scoreboard/all.** One further resource is **Broadband Now's** comprehensive compilation of ISPs in the United States: **https://broadbandnow.com/All-Providers.**

Gridlocate is a service that allows individuals and companies to use phone GPS trackers to follow enabled devices: **https://gridlocate.com.**

Though banned in the Apple and Google Play Store, the app **HelloSpy** can still be downloaded through the Internet. Intended for use on others' smartphones, it is advertised for parents, employers, and spouses with trust issues. It secretly records and logs SMS messages, call history, contact lists, photos, web history, and app-usage history once installed on a target's phone. Additionally, it records and logs the target's GPS information. Viewable at **http://hellospy.com/homepage.aspx,** HelloSpy is an example of the vast surveillance capabilities at the disposal of even ordinary citizens today.

Filings with the FCC by telecom companies complying with the E911 directive for smartphones: are available at **http://www.fcc.gov/ecfs/search/filings?sort=date_disseminated, DESC&proceedings_name=07-114.**

Chapter 3: Targeted Advertising and Consumer Privacy

Key Questions:

- What are modern methods of advertising?
- To what extent can modern advertising influence customers? Should there be limits to such influence?
- Can exposure to advertising ever be required of consumers of free or nearly free services? Under what conditions?
- Currently, who has access to the personal data used for advertising? Who should have access?
- How do consumers and corporations benefit from modern advertising? What dangers are there?
- Who is responsible for avoiding false or misleading advertising? The advertiser? The parties that enable display? Regulators? The consumer? To what extent is responsibility divided?
- What are current best practices for targeted advertising, and how often are they adhered to? Should there be legislation or industry action to improve best practices?

> Consider: Billy would like to purchase a new lawn mower. His old machine is well past its use-by date, and he is looking for an upgrade, even though he can't afford to buy it wholesale with his current savings and salary. Through his browsing habits, which include meticulous searches on hardware-store websites, his search engine seems to have noted his interest. Now, he is deluged with offers almost every time he opens his computer. Some of these advertisements, however, do lead to new models that he has not yet seen.

Additionally, Billy is fastidious about paying off any credit card debts. Several times a week, he monitors his account information online, opening up a new window and closing the browser immediately after. He also does enjoy seeing the numbers in his bank account, especially the spike during that first week of the month. Perhaps from these browsing habits, his search engine appears to have deduced the time of his monthly payday. As a result, he has started receiving loan information around then, and a recent loan offer has piqued his interest. It proclaims favorable terms and an apparently forgiving policy with payments expected but not required every month. Billy anticipates needing to perform various home-refurbishing projects in the coming months. As such, he is not sure that he will be able to repay the loan immediately, making this loan offer quite attractive. What he does not glean from the vocal advertisement is that while the initial two months are indeed easily forgiven with nominal payments, the rates steeply rise thereafter.

Did advertisers cross a line by even offering the loan to Billy? Did advertisers collect too much information by compiling and consolidating pieces of Billy's search histories?

—Inspired by the phenomenon of payday loans

"FOR SALE. 'Very Good Ground Coffee for Eight Shillings a Pound.'"

– *Boston News-Letter*, May 1761

The Boston News-Letter, one of the earliest newspapers in the United States, once printed the advertisement above, and many others of its ilk. In those early days, monopolies dominated the marketplace, competition limited. Additionally, with social prejudices overtly embodied in law, and natural geography yet to be shrunk via electronic communications, there were fewer target consumers. Hence, advertisement campaigns did not need to be particularly nuanced to succeed. They could, like the ad above, tell rather than show and did not have to motivate consumers to make the "right" choice.

Over time, a bustling marketplace has fostered the need for faster and easier differentiation. In a quest to maximize effectiveness, savvier marketers have moved away from wordy ads to pithy phrases and consistent graphics developed to root in the consumer's memory and incubate into spending. Corporate products now often turn on a single idea—nostalgia, a talking gecko, a jingle—although the core focus remains on implanting a brand name in the consumer's mind. While many such ads are chiefly intended to increase revenues, not all are exclusively so geared. Some companies advertise their sterling reputations and goods because they truly believe theirs is the best service, and that with some reasoning, you will see as they do.

Modern avenues of advertising are numerous and pervasive. They include telemarketing; e-mailing; flyer dispersal via physical mail; broadcasting via television and radio; printing in magazines and newspapers; product placement in TV shows, films, and video games; billboards and posters along highways and buildings; and online advertising via pop-up ads, banner ads, paid search placements, and content-disrupting ads. Many of these have long existed, but recent efforts have sought to capitalize on the enhanced connectivity of our age. Through the use of the Internet, from websites and phone apps, there are many more opportunities for us to receive advertisements. There are also many more avenues for us to reveal data about ourselves. The previous chapter explored several ways that we reveal locational data, and other information including browsing histories, network activity, and buying habits can also be tracked. Data miners now sort through that information to uncover correlations and identify the points of interest to prospective customers. From information about user habits and generalized patterns of behavior, advertisers can satisfy their fundamental craving to increase exposure of their products and ultimately sell them.

Opening the Cookie Jar

There are several prominent means by which information is collected on us on the web. Most commonly, data collection comes from third-party cookies, a type of marker initially designed to ease web surfing by minimizing the amount of data repeatedly reloaded. For instance, this stored data could include usernames, passwords, and site preferences. Such a cookie, stored when browsing web pages, consists of a few lines of code written by the advertiser that are attached to a website. These cookies may also include nuanced provisions such as a frequency cap to limit the number of ads displayed to a user or "burning" so that an ad disappears once clicked on or otherwise satisfied. Then, when someone visits the enabled site, the code drops an anonymized browser cookie upon the visitor, which stays with the user as he or she visits other sites, and repeatedly loads up the designated advertisement.[62] In addition to triggering ads on consumer browsers, these cookies can also more actively relay user information to the advertiser—information including sites visited and duration of visits.

There are other means of tracking user behaviors as well, including Evercookies, Flash cookies, canvas fingerprinting, and cookie

[62] Watson, "Ask the Decoder: Stalked by Socks."

syncing.[63] Whereas browsers typically allow users to clear or block cookies, these other trackers root more deeply, invisible to even specialized programs designed to minimize digital fingerprints such as the DoNotTrackMe browser plug-in or Lightbeam for Firefox. These more surreptitious trackers, employed by popular web pages, certainly raise a new set of concerns about privacy infringement, especially as they further penetrate the market.[64] However, for the purposes of this chapter, we will focus on the more standard goods.

Cookies are now everywhere on the Internet and used by even apparently innocuous sites like Dictionary.com.[65] In fact, in 2013, Mozilla policy leader Alex Fowler estimated that passing through just four popular websites meant exposure to ads from more than 120 different companies.[66] Nevertheless, cookie proliferation has long gone unchecked in the United States.

On smartphones, cookies are also employed for Internet browsers, but this is done with considerably more difficulty. The environment and different sorts of scripts used on phone systems, mean that ad display cannot be done as consistently as on laptops or desktops. Further, not all smartphone users access the Internet with browsers as frequently. Instead, advertisers circumvent the complications by incorporating ads and data trackers into apps, including the popular game Angry Birds (tracking location data as well as usage of other apps). Overall, mobile advertising has steadily increased in the past decade.[67]

---

[63] See, for instance, Schneier *Data and Goliath*, 48–9, and Angwin, "Meet the Online Tracking Device That is Virtually Impossible to Block" for more.

[64] Kats, "The End to Data as We Know It?"

[65] Angwin, "The Web's New Gold Mine: Your Secrets."

[66] Firefox, "Lightbeam for Firefox."

[67] eMarketer, "Total US Ad Spending to See Largest Increase Since 2004."

A recent joint study by Intel Labs, Penn State, and Duke found that publicly available cell phone apps are releasing consumers' private information to online advertisers, with perhaps 50 percent of apps emitting unique phone identifiers, while some even sent the phone number and SIM card serial number to developers.[68] Data leakage is hence not uncommon.

Who are the Data Brokers?

Though the primary means of data collection—and thus the potential impact of this practice—are relatively new, the advertising machine is long-standing. The data today feeds into a system in which third parties known as "data brokers" collect information about users that is then sold to other companies for use. Through these lists, companies can target interested individuals and save costs for themselves, while providing consumers with products of special interest. Through the use of such targeted ads, certain users like Billy may see payday-loan offers, whereas others will not.

Nevertheless, in compiling data to reach consumers, companies are moving well past the old regime that included data about a potential customer's age, race, occupation level, name, and address. Now, data brokers such as Experian also utilize cookie and query information to classify individuals by "life stage or lifestyle" and "life-event triggers" including marriage, childbirth, and home purchases.[69] Behemoth tech companies including Google are also entering the fray, and competitor Oracle has revealed its construction of so-called "super-profiles" where e-mail data, search queries, and data from third-party queries are all combined to craft a fuller profile for a particular user.[70] Recently, Silicon

---

[68] Enck et al, ""TaintDroid: An Information-Flow Tracking System."

[69] Experian, a "global information services" corporation, reveals some of these targeted list categories at Experian, "Targeted Consumer Mailing Lists."

[70] Roberts, "Google Faces New Pressure Over So-Called Super-Profiles."

Valley's poster child has taken some measures to curb collection, introducing a platform in June 2016 with which users could view some collected information, and a Dashboard in September 2017 that consolidated a user's sharing settings for perusal.[71] Additionally, in June 2017, perhaps in response to serious privacy challenges in the EU, Google announced that one vector of collection would be stopped.[72] Nonetheless, data collection as a whole widely persists.

As humans, we are skilled at ignoring the familiar and temporarily moving toward what is shiny and new. To that end, some consider personalized advertising a mere fad that will ultimately be proven inefficient and crumble away. Some studies have claimed to show that analysis using many variables, as the current framework allows, in fact only clouds the picture. To such critics, the modern market of targeted advertising entails a waste of resources, something they argue has led investors to dramatically reduce funding for Big Data analytic projects.[73]

However, there is something unique about personalized advertising; we humans also tend to be somewhat narcissistic,[74] so ads directly concerning us should continue to catch our interest. Furthermore, with the rise of AI and machine learning facilitated through cloud services,[75] personalizing ads is becoming both swifter and cheaper. Hence, it is highly unlikely that the current system will collapse

[71] Google, "My Activity" shows a user his prior activity and Google, "Google Dashboard" lets users view and control some of their data sharing settings.

[72] Gmail messages on all consumer accounts will no longer be scanned, but search histories and other avenues still will be, as reported in Statt, ""Google will stop scanning your Gmail messages to sell targeted ads."

[73] Taleb, "Beware the Big Errors of 'Big Data.'"

[74] For instance, see Mikulak, "Attention on the Brain."

[75] Chapters 4 and 8 will expand on machine learning capabilities and the role of the cloud in advertising.

from unsustainability. With the entry of large players, economic changes may restructure the landscape of data brokers, and thin out the number of competitors, this will not change the fundamental mechanisms of collection. In turn, reduced competition would only raise other concerns about possible monopolization of data collection and data services, to be discussed farther in Chapter 4. In any case, the basic fact of personalized advertising seems here to stay.

Guiding Questions

The case beginning this chapter is one instance of personalized advertising in the modern age. It is the product of personal data collection and dossier formation. Evaluating the appropriateness of the payday-loan offer will rest on a few crucial questions, highlighted from the "Key Questions" at this chapter's start:

> (1) To what extent can modern advertising influence customers?
> (2) Who is responsible for avoiding false or misleading advertising, and how is the responsibility divided?
> (3) Can exposure to advertising ever be a fair price to pay for free or nearly free services? Under what conditions?

Prospective answers to these questions will be worked into the differing perspectives below.

A Case Against Payday-Loan Advertising

The specific kind of loan proposed at the start of this chapter is a version of the "payday loan." These loans are typically short-term and small-sum, and through these, loan companies tend to directly and repeatedly access a consumer's checking account after acquiring initial user permission. While these loans are not always tied to one's payday, these lenders are typically based solely on the Internet with little-to-no contact information available. Insulated from inspection, it is little wonder that many of these lenders engage in fraudulent activities. Often, the hidden conditions of these payday loans tend to make the consumer's total costs exceed the original loan many times over, and in some instances, payment information entered by trusting consumers becomes the basis for identity theft.

Billy's case is moreover somewhat unique, because he has a regular bank account; payday lenders have historically targeted those who are underbanked with limited physical and credit access to a traditional bank account. Those who are underbanked are predominantly of African American or Hispanic descent, recent immigrants, or those with limited formal education.[76] Historically oppressed and often equipped with limited resources, targets may know no better about the scamming potential, and be driven by desperation with only limited options for other loans. With the current digital marketplace not adequately protecting consumers, it is in substantial need for overhaul with guidance to be found no farther than the European Union. Considering, as well, the high influence that modern personalized advertising may well have, personal data sharing as a whole should be revised.

Personalized Advertising: The Influencing Virus

There are degrees of influence that external factors can have on us. Opinions of those closer to us are likely to have more weight than opinions of strangers. Storms that occur in neighborhoods of special interest to us—places our family or friends live, for instance—are more likely to resonate than similar events in other areas. News stories about mathematical discoveries may intimidate us based on terrifying experiences in school, whereas stories about the annual Perseid meteor shower may instantly captivate us, in triggering a more romantic

[76] Burhouse and Osaki, "2011 FDIC National Survey of Unbanked and Underbanked Households."

memory. Cute puppies and kittens, however, are a common denominator, loved by all.

Whatever the reasons, each of us has particular interests. Advertisers try to meet those interests and through collection of personal data, they have more insight than ever before. At the very least, they can deduce correlations of value and repeated behavioral patterns. The question is, can advertising influence ever override a consumer's autonomy? If yes, then there is a deep problem with data collection as it occurs today.

Some studies have radically claimed that online advertising has no effect whatsoever.[77] However, most suggest the return on investment is substantial for advertisers. In fact, when the full thrust of advertising is considered, the case appears undeniable that advertising influences prospective customers. Importantly, clicking on an ad link isn't the only means of generating revenue—subliminally or subconsciously inserting a product or company name in a viewer's psyche can also induce future purchase.[78] Validating a precise causal process is difficult (if not impossible), but one attempt looks at the net revenue accrued by a company that buys such ads. A 2013 Facebook study suggested that 70 percent of their enabled advertisers "had a 3x or better return-on-investment" and half of the campaigns had a return five times their investment. These trends were noted even though less than 1 percent of Facebook users in the study actually clicked on displayed online ads. A subsequent study held that further personalized ads offered a projected improvement of 22 percent in returns for the near future.[79] Although

[77] For instance, see Blake, Nosko, and Tadelis, "Consumer Heterogeneity and Paid Search Effectiveness" and Terlep, Vranica, and Raice, "GM Says Facebook Ads Don't Pay Off."

[78] Overexposure can also, of course, actively cause a consumer to avoid a product out of irritation. However, even in such cases, the viewer retains the name recognition.

[79] Manjoo, "Facebook Followed You to the Supermarket."

Facebook may have a vested interest in demonstrating the success of advertisers on their platform, repeated demonstrations of persuadability and recorded revenue growth for participating advertising companies indicate the success is no fluke.

Of course, some of these studies might also be victim to "endogeneity bias," where users who click an ad were already interested in the product and would have purchased the item regardless. In such cases, successes will be over-represented. Nonetheless, while the segment of people already primed to purchase undoubtedly exists, the ads can still be effective on the second tier of those who were weakly considering purchase and then are pushed to actually do so. This competing consideration to endogeneity would also be difficult to accurately assess, and we could expect these to balance each other out.

Political Parallels

Similar advertising procedures to personalized advertising online have been used in political battles as candidates curry favor and seek to acquire voters. In his 2008 presidential campaign, President Obama engaged in a then much-ballyhooed data-centric approach, a process on which he improved in his reelection bid of 2012.

The Obama team's effort was multifaceted. In 2012, Obama's team had consolidated data into a single internal database that allowed the analysts to run tests predicting which people would be persuaded by certain kinds of appeals, with quantification of "persuadability" and information on race, neighborhood, and the expected impact of having a local volunteer reaching out rather than one from another state. The team further used an online get-out-the-vote effort with an app that sent downloaders pictures of friends, persuading the user to click a button and automatically message the friends to perform various voting actions. About 20 percent of those contacted through the app actually followed up on the request, contributing to a substantial number of votes.

In identifying whom to target, the campaign followed a similar strategy as previous campaigns that ran as far back as President

Eisenhower's 1952 election.[80] However, the Obama campaign diverged in also buying ads for unconventional popular programs such as *Sons of Anarchy* and *The Walking Dead* as it sought to strategically expand and re-energize the president's base.[81] Precisely establishing a causal connection between the data scientists' efforts and election outcome is a sketchy venture, as any number of influences may have contributed, including debate performances, international crises, and opponent blunders. However, by gross numbers, Obama won the 2012 election with 61.9 percent of the Electoral College and 51.1 percent of the popular vote. While his numbers were down from 2008 (67.8 percent of the Electoral College and 52.9 percent), they were still superior to the numbers of his less data-obsessed opponent and well-aligned with pre-election predictions.

The US presidential election of 2016, however, was quite a different story for the data-driven. Before 2016, Hillary Clinton inherited much of the Obama-era database, and she placed high value on the significant enhancements her team believed it had put together. Though similarly driven by numbers, according to insiders,[82] Clinton ended up losing with 42.7 percent of the Electoral College and 48.0 percent of the popular vote, in contrast to the 57.3 percent of the Electoral College and 45.9 percent of the popular vote won by opponent Donald Trump, to the surprise of nearly all pre-election data modelers.

Failure of the projected data to match reality could be the result of any number of influences. Even though numerous journalists have since sought to diagnose those influences, no decisive answers have yet emerged. The failure may have been the result of deliberate misinformation from those polled, unexpectedly significant events near

---

[80] Brownell, "This is How Presidential Campaign Ads First Got on TV."

[81] Scherer, "Inside the Secret World of the Data Crunchers Who Helped Obama Win."

[82] Goldmacher, "Hillary Clinton's 'Invisible Guiding Hand.'"

Election Day, outdated assumptions about Clinton's opponent, a new kind of candidate, and deliberate attempts to disenfranchise particular voters.[83] Moreover, while the data was one significant area for error, incomplete algorithms could also be held responsible,[84] undervaluing certain aspects such as the variability of Rust Belt states with respect to Democratic support. In a large sense, the 2016 US election seemed to unmask the folly of enormous data collection. At least, this has been the diagnosis of most talking heads.

Alternatively, the election might actually validate the power of data to influence. In contrast to Clinton's team, the Republican candidate initially used only an informal measure of "emotion," relying on natural charisma to reach potential supporters. However, late in the game, in the summer of 2016, the Trump team did employ a small data firm named Cambridge Analytica, which deduced demographic shifts from early returns and engaged in targeted advertising.[85] Incidentally, this was the same firm behind the successful Brexit campaign as well as other underdog campaigns that sang triumphant, including Ted Cruz's Republican primary bid in Iowa and various close congressional races.[86]

Though the firm has not publicly revealed many of its techniques, leaders of Leave.EU have offered limited insight. According to these sources, Cambridge Analytica was constantly monitoring "moods" on social media. When they sensed a trending topic and a chance to reach voters, they signaled to the campaign that they ought to jump in with a tweet or other message. Moreover, Cambridge Analytica also compiled data about personality, decision-making, and motivation,

---

[83] Vanian, "How Bad Polling Data Fooled Everyone Except Donald Trump" and Lapowsky, "Trump's Big Data Mind Explains How He Knew Trump Could Win" discuss these efforts further.

[84] Much more on algorithms will follow in chapter 4.

[85] Mezzofiore, "How a Little-Known Data Firm Helped Trump Become President."

[86] Resnick and Harris, "Will Brexit Masterminds Work for Trump?"

with characterizations including "openness," "conscientiousness," "extroversion," "agreeableness," and "neuroticism" determining individuals to target.[87] Incorporating such analysis gleaned from social media and other sources, Analytica sought to influence voters just as other candidates had tried in the past. Their data ultimately won the clash.

Empowered by modern technologies and compiled data, political candidates can market their messages to particular groups. Candidates from both dominant American parties have done this, and this enables politicians to exploit hope for the sake of personal profit—their power. Politicians have long borne the charge of being flip-floppers who tell you to read their lips, then break their promises in action. Many of them may in fact have good intentions, but without the light of transparency into their data operations, empty promises can be specially mass-produced.

Like politicians' secret databases, the dossiers of data brokers and tech giants have powerful potential. Altogether, the corporate and political numbers suggest that advertising can be effective, though there is nothing to decisively determine that advertising subsumes the individual's volition. Fundamentally, measuring influence is an impossible task; establishing causation, least of all via social-psychology studies, is a quixotic enterprise. Nonetheless, when there is enough smoke, there is almost certainly fire.

More importantly, the mere potential is enough to change the system. We cannot act only after thought control becomes a reality. If the right pieces of data can paint the picture of one's psychology and tendencies, then corporations with access to these moods will have enormous and unprecedented power. Absence of damning evidence thus far does not signify an evidence of absence; rather, it seems plausible that undue influence is only just around the corner. It must be controlled before it runs amok, and undermines the power of citizens.

---

[87] Ibid.

Brokers, Feeding on the Broke and Broken

Overlooking influence for the moment, the nature of at least some of the personal data compiled and sold is indubitably disturbing and dangerous. Payday loans are but one instance of an abusive practice. Despite the fact that they have been prosecuted in some cases,[88] they are not officially illegal—yet.

Many people are in dire straits: perhaps they have Alzheimer's, perhaps they are desperate owing to a terrible medical diagnosis or financial misfortune, or perhaps they are simply lonely and eager for any interaction. In any case, they may be described as one data broker executive did: "gullible. They want to believe that their luck can change."[89] Various data brokers easily traffic in lists composed entirely of such vulnerable people. For instance, InfoUSA sells lists with names like "Elderly Opportunity Seekers"; "Suffering Seniors," which contained 4.7 million names of those with cancer or Alzheimer's; and "Oldies but Goodies," which featured five hundred thousand gamblers over fifty-five years of age. For each of these lists, names were available to companies for 8.5 cents a head, and the lists were sold to abusers as well as repeat abusers at no direct cost to the company.[90]

Similarly, it is only a matter of time before the detailed data collected by corporations is analyzed to determine the pressure points of consumers. As Iowa Attorney General Steve St. Clair noted, "the attitude of [many of] these list brokers is that it's not their responsibility if someone else breaks the law."[91] A common check on corporations for good behavior is reputation—reputation of maleficence or wrongdoing

---

[88] Burhouse and Osaki, "2011 FDIC National Survey of Unbanked and Underbanked Households."

[89] Duhigg, "Bilking the Elderly, With a Corporate Assist."

[90] Ibid.

[91] Ibid.

will make consumers loath to invest in the company.[92] However, the data brokers at the heart of much advertising hierarchy are rarely known to the public; they are only beholden to the profits they bring their clients. Moreover, they are subject to lax regulations in much of the world, including the United States. Only a few revelations of private fiduciary information have been prosecuted, while most realms remain beyond reach.

Many organizations certainly have legitimate products to sell. However, payday loans are a detriment to the aggregate good of society. Payday loaners prey upon the unfortunate, using terms that signal financial distress such as "need money to pay rent"[93] to pounce. Moreover, the loans have been shown to statistically promote a great harm.[94]

The Way Forward: Transparency and Regulation

Outrage is naturally limited to what is visible, but it should also not be influenced solely by recent events. In the early 2000s, Target sent a teenager an ad booklet primarily featuring maternity goods based on her prior purchasing habits at the store and correlations those purchases had among the general population.[95] Through the work of a pioneering statistician named Andrew Pole, the company had identified the customer to likely be pregnant. After seeing the ads, her father angrily contacted customer service to demand an apology. He later learned, however, that his daughter indeed was pregnant. Since then, Target has worked to hide such targeted coupons deduced as valuable among

---

[92] Wells Fargo and Enron are recent examples of companies that have suffered for former unethical practices.

[93] Hauser, "Google to Ban All Payday Loan Ads."

[94] Burhouse and Osaki, "2011 FDIC National Survey of Unbanked and Underbanked Households."

[95] Hill, "How Target Figured Out a Teen Girl Was Pregnant Before Her Father Did."

generic ones. Yet, while there was substantial outrage at the time, the sentiment has since dwindled. Why is this?

Was it exposure of personal information to the girl's father that shocked people so—or was it the accurate inference made by the company? If the latter, then the problem must reside with the data collection itself. Today, even if web advertisers shield their particular product among others of a generic nature, the data is still being collected surreptitiously and compiled unashamedly with minimal consumer knowledge.

SPOTLIGHT: HOW CAN COOKIES BE CONTROLLED?

In a March 2012 report, the Federal Trade Commission (FTC), which regulates most privacy law in the United States, laid down particular privacy focuses for new technologies. The agency established five priorities, including increasing transparency into data brokers' collections on consumers; combating the massive tracking possible by large platform providers such as ISPs, browsers, and social media; promoting self-regulation; and increased transparency into data collected by mobile companies. Underlying each of these, and the chief action item highlighted in the report, was broadly establishing "Do-Not-Track" (FTC, "Protecting Consumer Privacy"). "Do-Not-Track" would provide consumers the option to not be tracked if they so choose, similar to existing EU legislation that requires users to affirmatively assent to cookie deployment on web pages. Yet, Do-Not-Track legislation has consistently failed in Congress and appears to have no way forward at the present. Hence, in actual practice this has left advertising an open hunting grounds for marketers.

For all of these reasons—the informativeness of high-resolution data, the danger of the substantial influence of such compiled data, the danger of compiled data's breach, and statistics strongly suggesting systematic abuse from particular targeted ads—the current advertising landscape, including the case of payday loans offered to Billy, is simply unacceptable. This is especially clear when particular services offer advertisements of a gentler, less invasive manner. Spotify, Netflix, and Amazon, for instance, offer suggestions for products based on user ranking input, and browser history. Certainly, their purview is limited, but they are still able to offer high satisfaction by relying on voluntarily submitted data.[96] As computer security professional Bruce Schneier notes, "Amazon's using your data in the same context it was collected, and it's completely transparent to the user."[97] The transparency in their subtle advertisements for other products is sorely lacking among many other sites and mechanisms. Unless action is taken—and taken quickly—even these companies may well modify their data-collection policies to remove their self-imposed restrictions and join the greedy fray.

The possibility of improving the landscape is also highlighted by policies across the pond. The EU, with its 2009 Cookie Directive and a 2017 proposal, has taken steps to legitimize cookie use, requiring users to actively assent to any browser cookie before deployment.[98] Cookies may be delicious in small quantities, but too many can be deadly; the rest of the world should note the EU's example.

While abusers can find ways in which to abuse any system, a poorly designed infrastructure only enables abuse. With limited protections currently in place and abuses spreading, systemic regulation for digital marketing is necessary. There are numerous options for such regulation. Regulation can follow closely from currently successful models that incorporate transparency, or from a comprehensive plan in the vein of the European GDPR. Alternatively, a unique American brand can emerge, finally actualizing those recommendations of the 2012 FTC report "Protecting Consumer Privacy in an Era of Rapid Change." The time for change is here.

---

[96] Amatriain and Basilico, "Netflix Recommendations: Beyond the 5 stars (Part 1)."

[97] Schneier *Data and Goliath*, 51.

[98] Kats, "The End to Data as We Know It?"

A Case For Preserving Payday Loans

In June 2016, Google decided to ban certain varieties of payday loans—those for which repayment was due within sixty days and the annual percentage rate was 36 percent or higher.[99] According to the company's research, such loans in particular led to financial harm in many cases. No other major companies, however, have followed suit. Still, Google's action was seen as a victory for numerous civil- and human-rights groups,[100] given that those often harmed by payday loans were those already in difficult circumstances.

While reducing some harm, the Google ban is a serious overstep. It overlooks the inherent value of payday loans when weighed against the volition of people, the necessity of revenue, and ultimately, the reality of monopolized power tied to the danger of censorship. For all these reasons, the ban actually commits more harm than good.

Maximizing Consumer Choice

First, architects of the Google ban on payday loans fail to appreciate that some individuals truly have no other loan options, whether from poor credit history or lack of access to traditional banks. In some cases, payday loans certainly have not been aboveboard or legitimate; however, this does not apply to all offers. Consumers have the volition to research and choose options that are reasonable to them. They also have a responsibility to parse what is true and what is not.

At the heart of this problem is again the question of volition: to what extent can advertising influence customers and can it subsume volition? Advertising has been around for centuries in one form or another, and scam artists have persisted through that time as well. In navigating past the scoundrels and false prophets, solid research has always been required, querying past clients and reading through expected

[99] Hauser, "Google to Ban All Payday Loan Ads."

[100] Graff, "An Update to Our AdWords Policy on Lending Products."

terms. Even this comparison, however, gives scamming lenders too much credit. Resources today are more ubiquitous than ever. A search engine and the bounties of the web literally lie at a user's fingertips, at the user's immediate disposal. When a consumer is presented with a biased perspective from a particular source, he or she still has the power to search for dissenting opinions supported by data.[101] Precluding access to a means of financial liquidity is wrong and subverts reason.

Moreover, the fear that collected personal data will provide windows into souls remains unsupported. The deductive power of data is clear from various cases such as the Target advertisement of the early noughties, but the advertiser's job is simply to put tantalizing options before the customer; the customer remains the one who chooses whether or not to act on the opportunity. In elections and for commerce, no evidence has yet seriously suggested advertisers can perform mind control. To claim that people could be programmed by simply pushing a few buttons insults their rational capabilities and invents an excuse.

It is true as well that the shadow operations of data brokers are troubling. As noted in a 2014 FTC Report, the operations of certain kinds of non-financial data brokers are opaque to the public and even to Congress.[102] Perhaps broader public awareness should be raised, and increased transparency shined upon the data broker industry. The advertising machine, itself, however, need not be drastically overhauled such that cookie acceptance becomes an onerous process.

Instead, individuals are endowed with responsibility. They may be influenced to a degree by others, their own environments, and advertisements, but they also have the power to overcome that influence,

---

[101] Of course, this excludes countries or areas under dictatorial regimes or with censored Internet, but thankfully much of the world is not under such a shadow; for such areas, more fundamental problems with privacy likely exist, and a situation such as Billy's is unlikely to materialize in the first place.

[102] Federal Trade Commission, "Data Brokers: A Call for Accountability and Transparency," i-ii.

as is evidenced by the many exceptions and exemplars through history who have defied unfortunate upbringings to benefit society. Fear of "mind control" is unfounded, and before rushing to overthrow or shackle the system, our immense individual power should be exercised.

SPOTLIGHT: CAN SPAM CALLS BE REGULATED?

Today, advertising from older means, such as phone calls, remains prevalent. The US Telemarketing Sales Rule (TSR), established in 1995, sought to provide private citizens with the ability to choose whether or not they received unsolicited sales calls. Among other moves, it restricted the hours telemarketers could call—from 9:00 a.m. to 8:00 p.m. in the local time zone—and threatened fines of up to $16,000 per violation. The TSR also required telemarketers to identify themselves at the beginning of any sales call and provide recipients the ability to opt out of future messages. In 2003, this opt-out function was transformed into the National Do Not Call Registry, a registry to which citizens could sign up and by which telemarketers would have to abide, with a few particular exceptions, on pain of greater financial penalty. The registry is perhaps the most popular program ever created by the FTC.

As you may have noticed, however, the effectiveness of the registry has dwindled, particularly since around 2010. As of May 2016, FTC estimates held that scam calls stole $350 million a year from unsuspecting consumers (Olen, "Congratulations! You Lost"). This is in part due to technology that has made it easier for callers to "spoof" or reroute their call, so that it appears to come from a nearby area code. Open-source software such as the prevalent VoIP has made the once-complex telephony process much simpler to abuse. Typically for such calls, the actual operators are located outside the United States in countries such as Jamaica or India, and the call is entirely a scam, designed to acquire consumer payment information for identity or bank theft. Just as other international laws have not taken deep root in the United States, US regulations have had difficulty in spreading beyond its own borders.

Understanding Pay-to-Play

Another important consideration too easily forgotten is that the corporations that provide services, including web browsers and social-media interfaces, require profits to operate. They provide an inherently valuable service in connecting the world, providing unprecedented access to information while ultimately opening doors. However, they must also be funded, and given that consumers have been notoriously loath to pay for Internet or other services (not to mention the practical impossibility of undoing public telecom and network projects), advertising seems a reasonable option. Moreover, advertising is not inherently bad: for corporations, there is an opportunity for profits via successful advertising; for advertisers and inventors, there is the opportunity to make profits in sales, as well as the satisfaction of drawing attention to potentially beneficial services or goods; and for consumers, there is the opportunity to obtain superior and useful products. Specialized medical devices, articles of interest, toys your friends will love, and more can be targeted at individuals who genuinely need them.

Without some monetary profit, corporations such as Google and Facebook cannot exist to provide their valuable services. The companies are naturally owed some payment for their otherwise free services, and advertising is a mutually beneficial option.

This does not mean, of course, that the present system is ideal or wholly acceptable. Consolidated personal data in databases remains a danger; hackers, stalkers, or terrorists might use location or other types of identifying data for their own purposes (see Chapters 2 and 8). Data lists of "Suffering Seniors" should likewise not be sold to just anyone (although there may be value in having such a list available to limited parties), and better techniques can be used to authenticate buyers. All of this can be done, however, without fundamentally changing the present advertising regime or imposing crippling regulations. Instead, all that may be needed is changing a few basic practices.

Self-Regulation to Combat Censorship in Monopoly

Another reason why a ban on particular advertising is dangerous stems from the monopolization of today's cyberspace. A ban on payday loans is simply part of a larger trend towards censorship. Whether the corporations involved intend such a shift or not, they should not be allowed to control content.

Today, a handful of large corporations have enormous control of key markets. For instance, as of 2014, Google controlled two-thirds of the US search-engine market, while Amazon controlled 30 percent of the US book market and 70 percent of its e-book sales.[103] One commentator has suggested that "the [modern tech] companies are analogous to feudal lords, and we are their vassals, peasants, and—on a bad day—serfs."[104] A key difference from the medieval chains, however, is that today's feudal lords are typically multi-headed, composed of a board, and hence, less prone to the whims and avarice of a single crazed individual.

Nevertheless, monopoly opens up avenues for censorship, an especially pertinent issue given that tech companies control more than simple commodities; they control our very means of accessing information. Even if justification for an initial act of censorship appears benign today, agreeing to control of information flow still sets a dangerous precedent. Prohibiting payday loan advertisements is just one wrongheaded instance of censorship.

Unfortunately, a recent history of increased regulation has substantiated such worries. In 2012, Google banned other material it judged offensive, including ads for gun sales, a ban that continues into the present. Beyond advertisements, the ban also affected the Shopping feature of the Google search engine, removing particular brands from

---

[103] Schneier, *Data and Goliath*, 57.

[104] Ibid., 58.

product results entirely.[105] Meanwhile, several lawsuits in Europe filed in November 2016 further portend something rotten, as they challenge the reliability of Google, accusing the company of tainting its services to promote its own brand.[106]

Furthermore, since 2016, a newfound obsession with "fake news" and apparently offensive video and textual content has gone viral. The infatuation even resulted in marketers pulling out of Google video advertising in record numbers, substantially harming profits. In turn, companies including Google and subsidiary YouTube, Instagram, Microsoft, and Facebook have all been exploring ways of hiding and deleting offensive content. Ostensibly, a focus of these companies is on combating terrorism, undeniably a noble goal, but other comments and content are also caught in the fire. In June 2017, Instagram launched a new artificial intelligence bot called DeepText trained on machine learning and intended to hide or eliminate hateful comments. The program inevitably has run into false positives, shunning benign messages, and the designers' goal is to be wrong only 1 percent of the time.[107] With more than 250 million daily Instagram users,[108] 1% of comments is still a substantial number. Efforts by Google, Microsoft, and Facebook are following similar lines, heavily incorporating machine learning to automatically flag and hide content without necessarily appreciating the context. Even more troubling, although the techniques

[105] Miniter, "Google Sadly Joins the Anti-Gun Brigade."

[106] Scott, "Google Rebuts Antitrust Claims in Europe."

[107] Thompson, N. "Instagram Unleashes an AI System to Blast Away Nasty Comments" discusses the Instagram project. Hughes, "Google bans adverts," Kastrenakes, "YouTube will hide, but not remove, some 'supremacist' videos," and Frier, "Facebook Moves to Block Ads on Fake News, Offensive Videos" discuss new bans by other tech giants.

[108] Constine, "Instagram Stories hits 250M daily users."

for censorship are being rolled out, there is no mechanism by which posters can challenge the corporate decision.

Allies of Google's anti-payday loan policy might argue that just as there are degrees of hate speech, not all censorship is the same. They might say that while omission is one form of censorship, deliberate misrepresentation is another that is at least as potent. As such, deliberate misrepresentations in advertising are violations of consumer trust and should be banned whenever discovered. By way of a tacit contract, consumers have a right to expect that corporations, including advertisers, will not deceive them.

Indeed, this principle of justice is the basis for most privacy law in the United States. Since the foundation of the Federal Trade Commission in 1914, under Article 5 of the FTC Act, America's primary privacy protector has worked to prosecute "unfair and deceptive" practices. It has targeted large companies including Google, Facebook, and Microsoft to set examples for security standards and consumer protection.

Nevertheless, the pro-prohibition argument is the beginning of a slippery slope. Once some information is prohibited for being offensive or dangerous, what will stop more restrictions? After all, it has been unreasonable ideas—increased civil liberties in colonial America, emancipation in the Civil War, rock 'n' roll, and the Internet, to name a few—that have ushered in innovation and progress. What seems unreasonable one day, could very well be acceptable the next. To compound the complexity, payday loans are not always bad: they can provide hopeful options for some of those targeted. Wresting autonomy from citizens is not the answer.

There is more to unpack regarding monopolization and censorship, and Chapter 5 will dive deeper into the related issues through the right to be forgotten. Even without wading further into the debate, there is a superior solution that coheres with citizen responsibility while combatting advertising abuse.

Rather than the censorship of blanket prohibition, a more promising approach would bolster transparency so that consumers can make more informed decisions themselves. After all, the nature of prosecution is that some abuse first occur. At all times then, even with regulatory protectors, the onus remains on us, as recipients of a service, to remain vigilant. Companies have repeatedly abused and even in the face of fines and public shaming, immediate profits sometimes blind long-term perspective. The history of deceit powerfully arms us with the wisdom to be wary. We have the power to examine data ourselves and in many cases verify a corporation's reputability.

Fortunately, there has been a recent trend of transparency among corporations, with transparency reports and public privacy notices published more frequently. Facebook, for instance, announced in 2014 that it would give users the ability to see the dossiers of "likes" and interests it keeps on them, in addition to the ability to change, add, or delete information in those files.[109] Similarly, Google recently opened its user profiles to the data subjects.[110] This transparency is vital. Algorithms inherently perform a sort of censorship by weighting results in accordance with particular pre-determined characteristics; for search engines determining access to information, transparency is vital. In advertising as well, opaque dossiers undermine the citizenry.

Still greater transparency is possible and where companies are hiding data, we can push for increased visibility. In addition to transparency, the principles of privacy by design and simplified consumer choice, as advocated in the 2012 FTC "Privacy Report," can be upheld within the industry. Importantly, this trend of transparency requires active accountability and vigilance on the part of consumers. Autonomy will be preserved, and as long as the spirit of transparency is

---

[109] Goel, "Facebook to Let Users Alter Their Ad Profiles."

[110] Google's Transparency Report and similar efforts by other corporations will be discussed more in the next two chapters on algorithms and the right to be forgotten, respectively.

maintained while censorship is combatted, much of the present advertising infrastructure will remain defensible.

In Conclusion

Data reveals much, and much is perhaps still to be revealed about data. Both those fundamentally opposed to advertising and those in favor of a largely unregulated landscape (as presides in the United States) can nonetheless agree that the status quo requires improvements, particularly with regard to transparency of disseminated personal information. Transparency also undergirds another fundamental area in the modern digital landscape: algorithms and corporate privacy, the topic of the Chapter 4.

Further Exploration

At the **FTC "Online Tracking" page**, the organization offers information to the public about online tracking mechanisms: **https://www.consumer.ftc.gov/articles/0042-online-tracking**.

For those interested in learning what data has been collected, **Cadie Thompson's June 2016 article at *Business Insider*** identifies various ways to check **what one of the tech giants** (Google) **has collected about you**: http://www.businessinsider.com/how-to-see-everything-google-knows-about-you-2016-6.

The **World Privacy Forum's "Data Brokers" portal** features news of the latest developments on data brokers across the world: **https://www.worldprivacyforum.org/category/data-brokers**.

Several **self-regulatory bodies for digital advertising** have published guidelines for best practices. Two of the leaders are **NAI** (http://www.networkadvertising.org/understanding-online-advertising/what-is-it) and the **Digital Advertising Alliance** (http://www.aboutads.info).

Chapter 4: Algorithm Transparency and Corporate Privacy

Key Questions:

- How prevalent is algorithm use today?
- What are major dangers of algorithm use with respect to private persons? What are algorithms' significant benefits for individuals, corporations, and society?
- How does copyright law impact algorithm disclosure for corporations? How much should we value corporate privacy?
- Under what conditions, if any, should corporations be required to disclose their algorithms?
- Should disclosure be legislated by the government or an external body, or are industry best practices sufficient?

> Consider: A fishing store in a quaint coastal town has long relied on tourists for revenue. While competitors may have cheaper prices, Phishing With You is locally renowned for its unique catch that consistently attracts larger and healthier game. Many have—and will—swear by Phishing's products, as demonstrated in numerous glowing testimonials exhibited on their store website.
>
> Recently, however, business has suffered, and store owner Maggie has a theory why. Whereas previously the fishing store's website was one of the top hits, in the past month, she has noticed her site drop down to the second page of results on the most popular online search engine. To her, this move is inexplicable, as Phishing's quality has not dropped. Neither have her

competitors, the large, established, department-store types, improved their products. Yet, these competitors are now the most popular hits.

Maggie contacted the search engine company to request an explanation, but the company only claimed it would look into it. Every day she awaits an answer, her business is significantly suffering.

How detailed an explanation, if any, is Maggie entitled to, and how soon?

—Loosely inspired by 2015 Google updates.

We live in a world of rankings. On sites such as OpenTable, Yelp, and Zagat, restaurants are ranked for their perceived quality of food, their service, and their ambience. Via US News, primary schools, universities, and graduate programs are ranked for a variety of characteristics. It is largely from its prominent inclusion in the American and British Film Institutes' Lists of All-Time Films that *Citizen Kane* has earned clout as one of the greatest motion pictures. In turn, some of these ranking systems are themselves ranked relative to the rest.

Algorithms power such rankings, taking in some inputs to produce an outward classification. In addition to rankings, these algorithms also play direct roles in modern society. They dictate our credit scores and produce the order of search results we see on the web. They run computers and help corporations decide how to manufacture the machines. In certain districts, they inform parole decisions as well as how police departments allocate resources. Some companies employ them to hire individuals. Algorithms are moreover used to calculate the likelihood of election and market results, and in turn, they can influence the moves made by corporations and individuals.

These algorithms provide the basis for many decisions in the world today. Their impact is substantial. But what exactly are they?

What are Algorithms, Really?

In computer science, algorithms are commonly compared to recipes. Certain data, or ingredients, go in, and a certain result—a delectable dish, one hopes—emerges. Like recipes, algorithms can be simple involving just a few ingredients and cooking techniques; they can also be more complex requiring several iterations and fine dicing. These algorithms can be used to produce evaluations for action or simply to model data sets.

As models, the algorithms can serve even larger purposes such as decoding a chaotic data whorl or providing predictions. They can perform these tasks through a variety of ways. For instance, algorithms may be constructed by supposed specialists who already suspect which blend of variables is significant. Alternatively, they may be built primarily by trial and error, through various combinations that balance expertise with uncertainty. More commonly, today, they may be chiefly built by other algorithms. That is, through machine learning, a computer will intake Big Data (see Chapter 7) consisting of a large and varied set of data variables that come in with high frequency. Based on chosen statistical techniques,[111] some portion of this data will be analyzed, and connections between variables identified. In turn, some model-associating variables are produced, and the model's validity will then be further tested on another portion of the data set. Ultimately, if satisfactory, the algorithm can be implemented for actual prediction.

As evaluations, algorithms produce the fodder for important decisions. Numerical values directly emerge from these algorithms, and these numbers can be interpreted via analysis, or another sorting algorithm can be used to make a decision. For instance, in determining whether to issue someone a credit card, a bank will look at several factors (e.g., prior loan history, current bank account amount, current employment situation, or credit history of a sponsor, if applicable) and assign the applicant a score for each relevant factor. Through the algorithm, some value will emerge, and if that value is above some preset threshold, the bank will continue with the application; on the other hand, if the score is insufficient, the bank will issue a rejection. Increasingly, banks, insurance companies, and other world designers are also incorporating nontraditional data such as locational history and Google search history to identify risks (see Chapter 3 for still other applications). For such applications, where data is collated from sources without consumer knowledge, risks to privacy are manifest.

Given the wild explosion of data today, algorithms help us sort through the expanse and identify what is relevant and informative—or

---

[111] There are a wide variety of techniques available, including random forest classification, naïve Bayes classification, and linear regressions. These are also often used in conjunction. For more on these, see Kotsiantis, Zaharakis, and Pintelas, "Machine Learning: A Review of Classification," pps. 159-90.

at least what the algorithm designers might deem to be so. Algorithm use has moreover become so ingrained in financial institutions and web use that they are undoubtedly here to stay. Nevertheless, cases such as the one that began this chapter indicate their negative potential. In the absence of transparency, various issues emerge, including the possibility of intentional or unintentional discrimination, as well as reinforcement of systemic biases. At the same time, some corporations such as search engine companies fundamentally rely on their algorithms. They cannot be wholly transparent, as this would endanger their profits and arguably their products' utility.

Thanks to search engines, we all have access to the Internet's bounty, but this access has also fostered a dependency. As browsers, we rely on search engine algorithms to provide us with knowledge for personal growth and world news that will in turn become data inputs for our personal convictions, political votes, and financial commitments. On the other side of the coin, businesses, such as Phishing With You, rely on search engine rankings to rank their services for users: these companies' exposure and hence profitability depend on search engine algorithms. Yet, many key algorithms remain opaque, including those employed as rating formulas at insurance companies and those utilized by search engines. This opacity has its detractors as well as its affirmers. On the one hand, the opacity enables abusive and arguably unjust policies; on the other, it serves to protect the intellectual property of corporations and perhaps benefit the populace as a whole. How should we decide to move forward?

A Case For Limited Algorithm Use and Transparency

The opacity of important algorithms used today is deeply disturbing, since it fosters discrimination on many levels. It not only helps to justify historic discrimination, but it also provides the basis for new prejudice. Moreover, these mechanisms for discrimination strike into the very heart of privacy and human dignity by dehumanizing the individual. Further still, opacity precludes some corporations from deriving just profits and allows particular entities to unfairly dominate. Ultimately, since the basis for opacity is deeply rooted in prevalent data techniques, the use of algorithms must be severely constrained.

Machine Learning Encourages Opacity

Algorithm use in many industries today is frequently referred to as taking place in a "black box."[112] There is little insight into what these algorithms are, which factors are deemed relevant, and what scores are ultimately sufficient to authorize important decisions. To some extent, corporations deliberately impose a curtain of secrecy to defend against theft by business competitors and customer abuse. However, an increasing number of algorithms are opaque without choice.

Such fundamental opacity arises from expansive use of machine-learning techniques. Through analysis of Big Data (see Chapter 7), correlations can be determined that could not be easily detected via human analysis, if at all. Although some sci-fi concoctions purporting massive underuse of our brains may suggest we can someday decode massive data stores ourselves,[113] in practice, the sheer volume of Big Data projects confounds human scrutiny, no matter how brilliant the analytic mind. For organic brains, it appears that there are limits for processing and decoding connections. A computational program, however, can easily run through large data sets and identify a serviceable model through mathematical techniques. In terms of broad usability, the most successful techniques must be adaptable; they cannot chart too

---

[112] For instance, Frank Pasquale's *The Black Box Society* (2015) and Mayer-Schonberger and Cukier's *Big Data* (2013) expound on the image.

[113] The short story "Johnny Mnemonic" as well as the film *Lucy*, promote the idea that physically and mentally there is vast untapped potential in the human brain. Up to 95% of the human brain may remain unused, and if activated could allow us to think faster and, of course, become lethal masters of the martial arts.

narrow an approach. As such, they employ randomization in the process and hence emergent algorithm can be different each time the program is run.

In machine learning, even the programmer will not necessarily know the algorithm that has been constructed. It is an open theoretical question whether such discovery is even possible, but the answer appears to be in the negative. Numerous attempts to uncover the internal logic of black boxes have thus far been unsuccessful.[114] Moreover, although discovery of the internal reasoning has not been decisively ruled out, traditional set theory with Gödel's incompleteness theorems suggest that discovery is not possible on any consistent basis. Inescapable opacity by design is dangerous for numerous reasons. One major reason is the necessity of visibility for stakeholders.

Stakeholders Require Visibility

Even without intent for shady dealings, opacity opens risk. In fact, the case that begins this chapter was based on a series of Google algorithm updates in 2015 that had adverse consequences for unsuspecting parties.

Google—and other search engine companies—periodically update their algorithms for a variety of reasons, including self-serving behavior by website developers to increase their site hits and revenues. In April 2015, the "mobilegeddon" search-algorithm update was released to favor mobile-friendly sites in page ranking. Since smartphone Internet usage was increasing, such a reevaluation would better reflect changing market dynamics. Additionally, with advance notice, corporations could be reasonably expected to adapt if necessary.

At least, this was the hope. Yet, a month later, the unannounced "Phantom" update resulted in precipitous falls for various sites

[114] For instance, Castelvecchi, "Can We Open the Black Box of AI?" features a representative failure.

overnight.[115] Google reserves the right to lower the ranking of sites if it believes them to have engaged in unfair site traffic to game the system. However, in these cases, Google delivers warning messages and the change for an appeal.[116] There was no such warning for the sites affected this time; the update had simply changed the ranking criteria without informing its dependents. Given that most users only observe the first few links for a search query, economic impact of a drop in search placement can be severe. Combined with society's increasing dependence on the Internet and particular search engines, sudden algorithm changes threaten the livelihood of small businesses like Maggie's. Customers and businesses have a great economic dependence on the web, and hence users deserve greater visibility into search-algorithm workings.

Opaque Algorithms Foster Discrimination

Algorithms can also be incorporated both to justify existing prejudicial policies as well as to implement new ones. Many of these instances of abuse—intentional or not—result from faulty assumptions about the meaning of Big Data, the data that typically feeds algorithms.

For instance, some have concluded that sheer volume of data will mask any biases. This conflation was one of the chief causes for failure in the case of Google's Flu Trends. The designers of this ambitious project which ran from 2008-2014 at Google, hypothesized that users searching for particular key words related to the flu ought to reveal the disease's rate of spread.[117] Yet, their premises were invalid, and after displaying their results as a public service (and publishing their

---

115 Levy, "Google's 'Phantom' Algorithm Update Hits Websites."

116 Waters, "Unrest Over Google's Secret Formula" with the ability to appeal at Google's "Reconsideration Requests" page.

117 Ginsberg et al., ""Detecting Influenza Epidemics Using Search Engine Query Data."

results for academic credit), they eventually pulled the service. As they later realized, and as others mercilessly mocked,[118] users search for flu-related terms as a precaution, not simply when sick. Observing data displaying a massive spike in the number of flu cases only served as positive reinforcement, inaccurately ballooning the statistics. In general, when researchers fail to perform appropriate data hygiene, it does not matter whether the data is big, lean, or sparse—it is ripe to be biased and inaccurate.

Moreover, other misguided analysts have conflated correlation—all that data analysis allows us to determine—with causation. The results have included attempts to justify racial and other forms of discrimination. For instance, African Americans have higher rates of incarceration in the United States than all other ethnicities. American history has not been kind to African Americans, from slavery through indentured servitude and separate-but-decidedly-unequal policies. The results have systematically oppressed and limited opportunities, and the effects linger to this day: the ACLU has found, for instance, that sentences imposed on black men in the federal system are nearly 20 percent longer than those for whites convicted of similar crimes; in identical circumstances, unequal results are clearly unjust.[119]

Systematic oppression is conquerable, but blind formulas that rely on historical data will not help to overcome these biases. In 2013, cybersecurity researcher Latanya Sweeney published a paper showing that online ads discriminated. Through the algorithms used by Google AdSense (which displays ads when users use Google), she found that searches for names typically assigned to blacks (e.g., DeShawn and Jermaine) produced ads related to police arrests between 81 to 95 percent of the time; by contrast, names more common to whites (e.g., Geoffrey and Emma) produced such ads between 0 and 60 percent of

---

[118] Lazer and Kennedy, "What We Can Learn from the Epic Failure of Google Flu Trends."

[119] O'Neil, *Weapons of Math Destruction*, 24.

the time.[120] Presumably, advertisers are not seeking to actively discriminate; they only aim to produce ads providing the greatest fit to the suspected user. Nonetheless, they are undoubtedly relying upon historical data—data that though accurate in itself is shaped by prejudicial circumstances. In blindly using such data, advertisers perpetuate previous modes of discrimination.

At some point, it seems to have been forgotten that statistics speak on a gross scale and may not apply to any individual. The reductionist mentality of turning an individual into data, however, is running dangerously amok. From credit scores to profit percentages to community rankings, it is too often forgotten that the individual is precisely that—an individual. Gross numbers should be a tool for decision-making; they should not make the decision itself, especially not through a black-box process. A blind reliance on what previous generations valued as important through these numerical evaluations restricts innovation and, most importantly, contributes to unjust discrimination.

Price Discrimination Must be Prevented

Some would say that not all discrimination is bad, and they would include among this acceptable set price discrimination. Price discrimination, the policy of charging different customers different prices, is not altogether banned in the United States. During the antitrust fervor of the early 1900s, discriminatory pricing was prohibited in the Robinson-Patman Act, provided certain provisions were in effect. To be illegal, a corporate policy would have to offer discriminatory prices to "equally situated distributors" for goods of "like grade and quality." Further, the items sold would have to be commodities or purchases: the act did not apply to services or leases. Moreover, the result of the practice

---

[120] Sweeney, "Discrimination in Online Ad Delivery."

would have to be reduction or likely injury to competition,[121] both of which can be difficult to prove legally.

This act has been the only federal law on price discrimination, and as the wording suggests, it does not entail a blanket prohibition of the practice. As a result, price discrimination persists in prominent spheres, including the hotel and airline industries. A variety of vendors exists for each of these services, including Expedia and Hotels.com, individual airline and hotel chains, and Google. These sites serve as focused search engines that advertise themselves as finding the best offers. However, the results are variable for any two individuals. These sites use dossiers of personal data compiled by data brokers and large tech companies, including search history and resident zip code. Incorporating the characteristics that the algorithm designer has determined as salient—ethnicity based on name, economic prosperity of a particular zip code, prior purchase history, and so on—the same services are priced at different rates.

For those who receive lower prices, differential pricing may seem agreeable, but the risk of abuse is simply too great. All companies begin with a target market they wish to penetrate—war veterans, teenagers with affluent backgrounds, enterprising females, and so on. Sometimes, they may also have particular markets they wish to avoid, such as African Americans, as many cigarette companies did in the 1950s, when company owners did not want their brand to be "sullied" by such a correlation, particularly when their dominant market was highly pro-segregation.[122] Similarly, corporations wishing to avoid other market segments such as Asian Americans, homosexuals, or conservatives could employ data compiled by brokers to raise costs for those "undesirables."

Besides intentional discrimination, there is also the possibility of faulty correlations leading to exorbitant and unjust pricings. A member

---

[121] Federal Trade Commision, "Price Discrimination: Robinson-Patman Violations."

[122] Bogart, *Artists, Advertising, and the Borders of Art.*

of a community that is affluent on average need not actually be affluent himself. As a result, such an individual might not be able to afford a marked-up plane ride. Once again, reductionist data threatens to lose sight of the individual. For travel and hotel businesses, the risk of such discrimination is bad enough. The effect is exacerbated, however, if the practice of price discrimination spreads further, as it seems to be doing with retailers. The cosmetics company 100% Pure, for instance, has admitted a clever use of coupons. Using the machine-learning algorithms of tech start-up Freshplum, Pure seeks to identify customers likely to leave their website without making a purchase. To these customers and these customers alone, Pure offers sizable coupons, which has in turn spiked revenue more than thirteen percent.[123] Differential pricing remains a legal practice.

For consumers, there is no visibility into how these discounts and different prices are offered. More importantly, owing to the potential for abuse, differential pricing is a practice that should be severely curtailed, if not stopped entirely. The practice makes a mockery of the free market and a customer's ability to choose, as corporations decide what the world pays. Even if different companies employ different algorithms to determine prices, the opacity of underlying algorithms wrests autonomy from the citizen.

Transparency Combats Monopoly

Monopolization further compounds the potential for abuse, and recent findings support the need for action against the current state of affairs. Google now possesses a massive market share. As of August 2016, it has 79.98 percent of the US search-engine market. Its closest competitor, Bing, has 9.78 percent, while the engine of choice for advertising despisers, DuckDuckGo, has 0.4 percent.[124] Google began as

123 Tanner, "Different Customers, Different Prices, Thanks To Big Data."

124 StatCounter, "Top 5 Search Engines in the United States, Aug. 2016."

a tiny company, and its growth is largely a testament to the strength of its product. However, monopolization is fundamentally dangerous. As Louis Brandeis commented in the early twentieth century with regard to trusts, "in a democratic society, the existence of large centers of private power is dangerous to the continuing vitality of a free people."[125] Even an organization publicly committed to not be evil will be tempted to waver in practice when it dominates the market. The public wields no recourse of economic pressure, since the dominator's competitors are eliminated. Moreover, without transparency into its affairs, to whom is Goliath accountable—and who can even determine that any abuses are occurring?

Trusts in the early twentieth century experimented with temporarily reduced pricing to squeeze out the competition and dominate the markets; thereafter, they raised prices to squeeze the public for profit. President Theodore Roosevelt wielded his big stick, and later, Presidents Wilson and Franklin D. Roosevelt worked to prevent these monopolies from wreaking too much damage. The Federal Trade Commission, America's leading privacy protector, was in fact originally designed to protect consumers by busting these trusts. Today, again, we are facing monopolies, though now in the technological arena. Among social networks, search engines, and online shopping sites, particular corporations have carved out massive chunks of the market. Even if sporting benign intentions at the present, each possesses the powerful tools for abuse.

Fear of market abuse is not simply a theoretical problem: there is some evidence that it has already occurred and will continue without regulatory action. Some have been highly suspicious that Google both favors its own products and punishes rivals. In 2010, specialized search engine Technorati had its search ranking diminished for what Google allegedly deemed an illicit attempt to game greater site traffic, and Technorati's founder suspected Google was retaliating against a market

[125] Taplin, "Is It Time to Break Up Google?"

competitor.[126] Where a corporation is infused with power to act as judge and jury and oversight is nearly nonexistent, we cannot decisively know the behemoth's motivations.

SPOTLIGHT: HOW CAN ALGORITHMS REINFORCE STEREOTYPES?

Besides Latanya Sweeney, several others have also highlighted algorithms' proclivity for prejudice. Claire Cain Miller's 2015 *New York Times* article, "When Algorithms Discriminate," points to more relevant studies. One jointly conducted at the University of Washington and University of Maryland examined the effects of exposure to biased data. The researchers found that among Google Image results for "CEO," 11 percent depicted women, compared to the 27 percent in reality. Moreover, a small but sizable portion of study participants (7 percent) were influenced by the gender gap presented in image results (Miller, C.C., "When Algorithms Discriminate").

The study shows that bias can feed itself in a feedback loop. This applies to prominent but poorly vetted tweets and posts as well as image searches. In late 2016, the same sort of feedback fueled the "fake news" crisis on Facebook, which continues to be a danger.

The underlying question is what reality should advertisements reflect? Should they reflect reality as it truly is in the present, an idealized reality as they see fit, or anything they desire? If they are to reflect reality, how can they ascertain it?

Google has made some efforts in the years since to make its algorithms more transparent. Several papers and explicit disclosures[127]

[126] Waters, "Unrest Over Google's Secret Formula."

[127] Google, "How Search Algorithms Work."

have elucidated a clear majority of the two hundred or so variables that apparently go into a Google search. However, this is a far cry from truly providing visibility into the algorithm's usage, as subsequent updates have not been disclosed, and neither is the weight given to each variable. As a result, several European countries have rightfully pushed for full disclosure to a panel of experts.[128]

Further fuel to the fire comes from a recent ruling that Google favorably advertises its own products at the expense of higher and equally-rated products. Following a seven-year investigation, a June 2017 ruling in the European Union fined Google a hefty $2.7 billion for its sly placement of shopping ads.[129] Calculated as up to 30 percent of Google Shopping revenues multiplied by the years of abuse, this fine was presented since "even the most highly ranked rival service appears on average only on page four of Google's search results."[130] As Google appeals, other pending suits may reveal still more abuse including whether the company has banned competitors in still other ways, such as prohibiting co-habitation on websites that incorporate a Google search bar or Google ads.[131] Already, it would seem, the search engine is falling prey to a mentality of maximal profits at the expense of accurate results.

This is not meant as an indictment of Google in particular; rather, it should illustrate the peril of a market dominated by any corporation. A lack of competition in the marketplace damages the value of services available. Price discrimination was originally banned in the United States when it was used with the intent of destroying all the

---

[128] See Lomas, "French Senate Backs Bid to Force Google to Disclose Search Algorithm Workings" and Vasgar, "Berlin Pushes Google to Reveal Search Engine Formula."

[129] De Vynck, "These Search Results Show Why Europe Is Mad at Google."

[130] Toplensky, "EU Fines Google €2.4bn over Abuse of Search Dominance."

[131] Boffey, "Google 'faces €1bn-plus fine' from EU over market dominance."

competition. The opacity of search algorithms cannot be allowed, lest we fall prey to a new monopoly.

For all these risks and the observed dangers of algorithms, transparency must be upheld. Moreover, better regulation is required, whether it be by governments, the industry, or a public firm.

One Solution: "Algorithmists"

The precise shape that transparency takes will have to be carefully conceived. On a constructive note—and there is much in algorithm use to cherish—authors Meyer-Schonberger and Cukier have intriguingly proposed a new profession known as the "algorithmist." Members of this group would take a vow of impartiality and confidentiality, much as accountants and certain other professionals do now. Professionals within and outside of companies, they would verify that corporate policies incorporate respect for the individual beyond the aggregate. They would also verify that corporate policies adhere to ethical and privacy standards.

Specifically, as proposed, these algorithmists would evaluate corporate selection of data sources, choice of analytical and predictive tools (including algorithms and other quantitative models), and the interpretation of results. In the event of a dispute, they would have access to the algorithms, statistical approaches, and data sets that produced a given decision.[132] In short, algorithmists would serve as a knowledgeable and trustworthy intermediary, providing a significant check upon corporate policies while also defending companies' intellectual property from easy poaching.

Of course, it is not a perfect solution. Critics might still claim that introducing a process of verification for every algorithm would delay and eventually stifle innovation. Moreover, the value of the verification process would hinge on the comprehensive quality of the particular

[132] Mayer-Schönberger and Cukier, *Big Data: A Revolution that Will Transform How We Live, Work, and Think,* 180.

ethical and privacy standards adopted. Further still, an unjust algorithm, especially when it incorporates a bevy of variables, cannot always be easily detected; iniquity might only be revealed through application. All of these complicate the picture, but are kinks that can be smoothed over as specifics are produced. An evolving set of evaluations, for instance, might be required to accommodate the variety of ways in which algorithms are used in different industries. Ultimately, an institution of algorithmists or other regulators could help realize a more just environment.

Another approach in parallel would involve composing clearer industry standards for data collection and analysis. Critical questions about data in both collection and analysis should reduce the number of unjust algorithms and resultant decisions.

Fundamentally, transparency must be upheld. At its inception, the Internet was seen as a tool for the democratization of knowledge, since any individual could access hundreds if not thousands of different articles, books, and news stories online. Through widespread access, citizens of the world would be empowered to learn and better make decisions. Today, however, corporate secrecy threatens that goal, with googols of sources controlled by unknowable overlords. Though necessary for users to utilize the digital libraries, these opaque algorithms also severely threaten the Internet's equalizing power. As such, these and other important algorithms require greater regulation and transparency.

A Case For Bolstering Intellectual Privacy

As even a critic of modern algorithm use will note, algorithms are invaluable today. The current state of minimal regulation with limited transparency is also essential to ensuring growth and retention of algorithms' benefits. These benefits manifest in several ways, from defense of trade secrets to enabling corporations to better spread the fruits of their labors. Additionally, opacity offers an essential defense against another sort of abuse by rogue users. Given that regulations can inhibit innovation and the discovery of creative, effective solutions to abuses, neither opacity nor budding policies of price discrimination ought to be discarded. The status quo ultimately upholds the power of the individual.

Opacity Defends Trade Secrets

Corporations may be "people" too. Such was the controversial ruling absorbed into US law with Citizens United in 2010.[133] The metaphor of many into one has a long and rich lineage,[134] and is essential to political theory. In republics such as the United States, the many are incorporated into the body politic via a constitution, with benefits to the body intending to benefit all or nearly all within. The extent to which a modern corporation can be said to exhibit such unity and warrant similar respect is certainly a contentious subject for consideration. Less controversial, however, is a corporation's fundamental right to defend its trade secrets.

For corporations, maintaining trade secrets precludes the competition from copying essential procedures and cannibalizing the author. Where there is a free market, corporations have an interest in staying ahead of the competition by protecting their products and unique techniques. This is clearly illustrated in the automobile industry, among numerous others, where engineers who discover a more efficient engine or vehicle design have a right to protect their intellectual property via copyright laws.

In Brandeis and Warren's articulation of the right to privacy, copyright law in fact served as their model for privacy. A copyright is a slightly different designator than a trade secret, since the latter is inherently protected as confidential, but both are types of intellectual property. Copyright law has recently expanded in many areas, especially

[133] Dunbar, "The 'Citizens United' Decision and Why it Matters."

[134] As discussed in de Pizan, *The Book of the Body Politic,* xviii, its origins stretch back at least as far back as ancient Rome.

digital technologies, as patents and ownership in cyberspace are less definitive. All the while, the interconnected nature of the world has seen an increase in global theft. It has become something of a rallying cry for floundering politicians, but there are also numbers to back up charges against deranged dens of copycats. In 2017, Congress's bipartisan Commission on the Theft of American Intellectual Property found that stolen trade secrets, pirated software, and counterfeiting cost the United States between $225 and $600 billion dollars annually.[135] The commission also found that counterfeit pirated tangible goods amounted to at least $143 billion in 2015, and at least 20 percent of these displaced legitimate US sales.[136]

Recovery of stolen property is difficult because international law is nonbinding. If a country such as China refuses to persecute homegrown copyright violations, there is little the affected nation can do to ensure justice. War is unreasonable, while trade embargoes only have a delayed effect, if any at all; the nature of international blocs requires widespread cooperation, and this is often hard to come by. Hardware manifestations of pirated software can be captured by customs officers at international checkpoints, but the amount of stolen material recovered there is a drop in the bucket. Ultimately, theft becomes damaging leakage that keeps flows ever faster.

Cyberespionage occurs domestically as well. Cases in the last few years have not unreasonably claimed theft of AI technology, including a March suit by Alphabet against Uber for self-driving cars.[137] Such cases tend to fall apart, however; it is quite difficult to prove theft. In some cases, similar inventions may in fact stem from thinkers operating at the

---

[135] Hvistendahl, "China's Theft of US Trade Secrets under Scrutiny."

[136] The National Bureau of Asian Research, "Update to the IP Commission Report 2017."

[137] Reuters, "Alphabet Adds Patent Claim to Uber Intellectual Property Theft Lawsuit."

front lines of technology, simultaneous inventors like Leibniz and Newton. Nonetheless, in our cutthroat business economy, companies struggle to maintain an edge, and it is not surprising that notions of fair play are sometimes thrown out the window.

With thievery rampant, a company's livelihood cannot be exposed for all the world to exploit. For a search engine company, page-rank algorithms are a critical trade secret, and adequate protection must be granted to them. While corporate abuse is a real concern, Google has made serious efforts to increase transparency about its algorithm use and other activities. Its annual transparency report stands as a shining example to other corporations.[138] Moreover, it has made a serious effort to help dependent businesses such as Maggie guidelines, with published general guidelines to maximize ranking.[139] Google's and other corporations' trade secrets should not be wholly exposed. In another respect as well—America's reliance upon a free-market economy—some degree of opacity is necessary

Corporate Profits Benefit Society

Some critics, including the EU courts, criticize technology behemoths for their monopolization and chastise corporations for prioritizing their own products. What is too often forgotten in this tirade is how technology corporations, with their algorithms, essentially provide their services for free. Certainly, they gain revenue from other sources, including advertising, that merit a closer look (see Chapter 3). However, in terms of the products themselves, the fruits are freely accessible to users.

In enabling a symbiotic relationship between user and product, search engines are unique creatures. Through the bounties of the web, organized by search engines' finely tuned algorithms, information is

---

[138] Google, "Transparency Report."

[139] Google, "Webmaster Guidelines."

widely available. Google receives the benefits of traffic and ad revenue, while global users perform more than forty thousand search queries per second to learn anything they want.[140] The now-legendary Library of Alexandria held a small fraction of the Internet's size in content; we should not make the same mistake of burning down our modern athenaeum.

It benefits the globe for the Internet to have skilled and willing caretakers, such as the current search-engine behemoths. Were Google's algorithms to be made more public, the risk of theft and abuse would not only impact the company but ultimately all its previous users. Moreover, Google has remained extremely responsive to consumers, indicating that it still recognizes its dependence upon users. Its now-standard publication of transparency reports, its increasingly user-friendly visualizations in response to feedback, and its expansion of services to include better language translation, journal access, and map content all testify to this commitment.

In the case of the recent EU fines against apparent favoritism toward its own products, the requirement that Google modify its ad interface within ninety days is blatantly unfair and unrealistic; experts expect the change to require at least several months, if not years.[141] As the unreasonable decree suggests, the EU are ulterior motivations, and a history of attacks upon American tech companies in the past ten years bolsters this claim. In 2008, the EU charged Microsoft with failure to comply with server interoperability;[142] in 2009, it fined Intel on a similar charge of market abuse,[143] and, also in 2015, it began seriously

[140] Internet Live Stats, "Google Search Statistics."

[141] Toplensky, "EU Fines Google €2.4bn over Abuse of Search Dominance."

[142] Free Software Foundation Europe, "European Commision vs. Microsoft: Chronology of the Case."

[143] European Commision, "The Intel Antitrust Case."

investigating Facebook's ad policies.[144] All were American firms—as most of the leading tech giants are—and various commentators have plausibly suggested that this animosity toward American tech companies may be ideological and social, a conservative bias toward the old and resistance toward the pioneers.[145] In the past century, Europe has not embraced the spirit of innovation in quite the same way as America. While a lack of regulation in America has allowed innovation to run amok to open technology's benefits, European law has focused on fear, preventing the dangers from developing.

All of this is to say that what European courts found objectionable is hardly damning. Based on precedent and intent, American courts are right to be more forgiving. Google's apparent coloring of the market at the expense of competitors is potentially dangerous—if it is indicative of a larger trend toward slanted information. To that extent, users and watchdog groups are right to always remain vigilant; however, this incident and the resulting fine should also not be blown beyond proportion.

Google is not an angel in white, but it ultimately offers a great deal more good than bad. Nonetheless, if all these reasons are insufficient, there remains the option to use or design a more desirable engine, such as DuckDuckGo or Microsoft's Bing, among others. They function well enough, even if their functionality is not nearly as strong as that of the market dominator. A revolution begins with one, and today, even amid market dominance, an individual can alter the entire terrain. This is one powerful version of the American dream, one that various tech company founders have achieved and one that remains open to all those who believe.

---

[144] Kottasova, "Europe Takes on Apple, Facebook, Google and Amazon."

[145] Bird, "Why Europe Hates American Tech Companies," Mirani, "Does Europe Hate Google or Hate America?," and Weinstein, "Why Does Europe Hate Silicon Valley?" are a few among many.

Opacity Blocks Bombs

Corporate privacy is also important in a grander way: defense of the public good. "Google bombing," deployed in the early twenty-first century, was a method by which users could repeatedly enter search words and click links to bring popular searches in line with their personal proclivities. Google's autocomplete search feature would then prominently display these manufactured searches. Infamously, through user exploitation, the term *miserable failure* was linked to President George W. Bush and later to filmmaker Michael Moore.[146] Many of these cases are more amusing than dangerous, but a variant of "bombing" allows for determined individuals to repeatedly click on some search query (perhaps their own business website) and, in so doing, raise that link's ranking. This would come, of course, at the expense of honest users and legitimately operating businesses.

To combat such selfish procedures, Google has modified its search algorithm so that it considers more than the number of times a link is clicked. Even with a more complex formula, nefarious users who discover the latest algorithm can try to game the system, and hence Google frequently performs updates. Furthermore, it maintains some opacity. Since its early days, Google has also incorporated other tests to determine whether sites are directing bots and other inconsequential traffic with the effect of maximizing the site's footprint. Google notifies such abusive sites, with a chance for appeal, [147] and punishes repeat offenders.

---

[146] Jackson, "10 Notorious Google Bombs."

[147] "Reconsideration Requests" can be filed at https://support.google.com/webmasters/answer/35843?hl=en.

Means of gaming the system still remain today, including "black hat" and other search-optimization techniques.[148] Yet, by modifying its algorithms and hiding some details from the public-at-large, Google seems to have succeeded at reducing the toxic bite. Self-regulation, rather than something externally imposed, best allows for abuses to be curtailed.

SPOTLIGHT: WHAT ARE OTHER WAYS ALGORITHMS CAN BE USED FOR ILL?

The discussion here has focused on devious efforts to raise a web page's ranking. However, the same end result can alternatively be achieved by sinking competitors' pages. There are a variety of ways users can perform negative search-engine optimization for competitors, including filing a false copyright complaint (which will lead to a mandatory, though temporary, removal) and duplicating a competitor's web site to divide its rankings hits. Another especially devious way is "search-engine bowling," where the deviant hires out a link farm to illicitly raise the number of hits that a competitor receives (e.g., Greenberg, "Saboteurs of Search"). As a result of the suspicious activity, after a temporary ranking boost, the targeted website is liable to be punished by search engines, and its cries of innocence will go unheeded.

Such deviousness can be seen as reason for either more transparency into search algorithms (so that webmasters can better defend against attack) or else less (so that adversaries do not learn which holes to exploit). In any case, these are still more reasons to pay search algorithms greater heed.

[148] See, for instance, Pasquale, "Internet Nondiscrimination Principles: Commercial Ethics for Carriers and Search Engines," 283, for an enumeration of various ways to subvert search engines.

Regulations Inhibit Innovation and Solutions

Externally imposed regulations and oversight are rarely well conceived. With fears magnified and arbitrary procedures installed, government procedures by the uninformed tend to limit progress and innovation. In some cases, such as the 2003 password guidelines recommended by NIST that grew pervasive yet were ultimately mistaken, regulation can actually undermine security.[149] In turn, these limitations also tend to limit the scope of industry-led solutions.

Solutions to stubborn security problems and privacy risks require collaboration and forward thinking. They are far more likely to come from professionals with expertise in a field than from bureaucratic neophytes, no matter how brilliant those novices may be. Hence, if regulation is necessary, it should come from within the industry, not without. These professionals are likely to have an appropriate understanding of which restrictions may be truly necessary as well as which restrictions are not.

To this end, a few different proposals for fair use are out there. The World Wide Web Consortium (W3C), consisting of volunteer professionals in information technology, has developed a series of technical standards for web use.[150] These standards for best practices are not universally agreed upon, but various proposals to bolster both security and transparency do float out in the beyond.[151] Proactively, others such as those at Search Engine Watch, have projected future trends of search-engine design. They also continue to monitor whether the risks and benefits materialize.[152] If more regulation is needed, it

---

[149] The NIST guidelines were replaced in 2017, as described in McMillan, "The Man Who Wrote Those Password Rules Has a New Tip."

[150] W3C, "Standards."

[151] See, for instance, Nelson, "Industrywide Standards for Document-Level Security in Enterprise Search."

[152] Boyd, "What Will the Future of Google Search Results Pages Look Like?"

should ultimately continue to come from within the industry, where it is likely to interfere least with progress while permitting beneficial development.

Price Discrimination Has Value

Discrimination comes from the Latin *discriminationem*, meaning "the making of distinction."[153] Historically, distinction-making has been a useful process and remains so today. It is only by discrimination that options can be evaluated as better or worse and decisions of action can be made. It is discrimination that allows us to select the ripest apple from a tree, enjoy our favorite genre of film, and spend time with those we truly love. The racially prejudiced do discriminate as well, but their basis is fundamentally unsound. It is that faulty basis which is wrong, not discrimination itself.

Price discrimination is a species of discrimination that is not inherently bad. What can be bad are inappropriate uses of data. As the critic of corporate privacy noted above, the real problem with certain price discrimination policies is their faulty data. Correlating race with propensity to crime, for instance, subverts the individual's autonomy and undermines the respect each individual deserves. And whenever such badly discriminatory practices are caught, the FTC will punish wrongdoers, as it has in past publicized cases.

Algorithm use fueled by Big Data is certainly not above reproach, but appropriate hygiene performed on data should yield better results. In the countless success stories of Big Data (see Chapter 7), keen vetting techniques have let to appropriate deductions and beneficial policies. For instance, a study of mobility in Turkey found a correlation between economic stability and low rates of displacement.[154] By analyzing activity on the ground and interviewing several of the

---

[153] Online Etymology Dictionary, "Discrimination."

[154] Singh, Bozkaya, and Pentland, "Money Walks."

participating residents, researchers concluded that the itinerants were not naturally irresponsible; rather, they were exerting energy in order to find new employment. As the researchers rightfully noted, however, even this explanation was not definitive. On an individual scale, these trends could break down.

SPOTLIGHT: HOW CAN ALGORITHMS REGULATE HATE SPEECH?

Algorithms are inherently neither good nor bad. Application makes them so. In mid-2016, Yahoo announced a tone-detecting algorithm that could detect abusive online comments in 90 percent of test cases (Reynolds, "Yahoo's Anti-Abuse AI Can Hunt Out Even the Most Devious Online Trolls"). Although significantly better than similar models by competitors, a 10 percent false positives and false negatives is still high.

In a world where anonymous profiles are employed for cyberbullying and fomentation of violence, hate speech is a grave danger that has driven some to suicide and others to mindless aggression. Lacking means to restrict such destructive discourse, various community forums have opted to restrict dialogue altogether. For example, news outlet NPR closed its public comments section in 2016, and film database IMDb shut down its discussion boards in February 2017. The expanding Internet is adding fuel to a long-standing fire: how should hate speech be regulated?

Respect for the individual is not distinct from appropriate algorithm use. Algorithms, however, do allow for increased efficiency. As the term *machine learning* suggests, there will be a learning curve for machines, but eventually, these programs and the algorithms they deliver can provide grossly accurate trends. Via more personal data (see Chapter 3), they can also provide personalized recommendations and products. Taken to its limit, price discrimination can offer the ultimate personalized experience, learning from a single consumer's behavior what he or she enjoys and what prices he or she is willing to pay.

Paying minimal prices is desirable in the abstract, but not sustainable or fair to the service suppliers. A market differentially priced—where some pay greater costs than others—may in fact promote a form of justice. If those with the means pay a greater cost for a service such as airfare, those with less can also pay more affordable rates. This stratification is essentially the layered tax model. Those who earn more typically pay greater amounts in federal taxes, whereas those earning less contribute less. The extra that those capable provide—extra that they can, at least ideally, live without—allows for the rest of society to benefit.

Importantly as well, as the Robinson-Patman Act has highlighted, price discrimination only applies to non-commodities. As such, users always have the choice of not purchasing an item which is exorbitantly priced. Commodities are a class open to some interpretation, and definition can be precisely scoped out and legally enshrined in different ways. Nonetheless, the cases in which price discrimination has so far been allowed, including purchase of cosmetics, are quite positively nonessential ones. Standard bread and milk, water, and other essentials need not be offered at differentiated prices, and there is no reason to think they will.

Moreover, if this view of a personalized world remains upsetting, solutions that encourage growth and production are possible. For instance, an alternate solution to a wholly customized marketplace would be developing a robust marketplace in which consumers can opt in and where prices are served at a flat rate. Of course, if the vast majority of society migrates to this latter marketplace, then an undifferentiated market will reign in the end. Ultimately, the people will have to choose what sort of economy they desire. The benefits of personalization suggest that differential pricing, fueled by algorithm calculations from personal data, will continue to have a rich future.

Algorithms serve to educate and innovate, and they allow for essential services to be swiftly performed. A certain level of opacity is necessary so that these services can be optimally produced. At the same time, citizens must actively continue to hold corporations accountable. Corporations are generally aware that they serve the public. Wherever they falter, corrective measures can be implemented. However, this should be done in a way that fosters growth and does not place unjust shackles upon innovation in the tech industry, which as a whole is socially beneficial.

At the heart of any changes should be respect for the individual. With data analytics that respects the individual and abides by best practices, citizens can continue to assert their autonomy. They can profit from algorithm use and policies such as price discrimination in a way that also builds a more equitable future.

In Conclusion

Privacy with regard to algorithms applies differently to both consumer and corporation. From each vantage, there are some different concerns and possibly differing evaluations of transparency's value. Protecting trade secrets, preventing discrimination, empowering stakeholders, and appropriately employing the use of price discrimination will all be important considerations.

In any case, the concerns of privacy and transparency extend well beyond any single company or application. Already, the impact of algorithms is vast—they affect what services we use, how we entertain, what we eat and consume, and in short, how we live. As technology's domain and potential balloon through the tools of Big and Small Data, the significance of this balance will only grow. It will be important to understand and stay aware of these issues. For Maggie and all of us, a full accounting of algorithms will also require an understanding of the advertising landscape today, as discussed in Chapter 3. Also relevant will be the lifespan of personal data out on the web, the topic of the next chapter: the right to be forgotten.

Further Exploration

The **FBI's Intellectual Property Theft/Piracy** news page contains information on the latest cases pursued: **https://www.fbi.gov/investigate/white-collar-crime/piracy-ip-theft/intellectual-property-theftpiracy-news**.

For websites that Google deems to have violated fair practices by illicitly trying to bolster rank algorithm hits, the tech giant typically sends a warning notice. Webmasters have the ability to appeal at the **Reconsideration Requests** page at **https://support.google.com/webmasters/answer/35843?hl=en**

Frank Pasquale's *The Black Box Society* (2015) discusses the vast extent of black boxes in technology and especially in finance today.

Viktor Mayer-Schonberger and Kenneth Cukier's *Big Data: A Revolution that Will Transform How We Live, Work, and Think* (2013) focuses on the impact of Big Data upon the world and how black boxes contribute to that revolution.

Chapter 5: Deletion Technologies and The Right to Be Forgotten

Key Questions:

- Does there exist a right to be forgotten? If so, of whom and by whom?
- What information can be "forgotten" in this context? What information, if any, should always remain?
- How do we balance the tension between a right to be forgotten and a right to free expression?
- Do today's technology and infrastructure support a right to be forgotten?
- Is it possible to "forget" information in only particular regions or nations? If so, what would be the impact?

> Consider: James, a former mathematics child prodigy, experienced a terrible tragedy and grew disenchanted with fame as he grew older. He sought a quieter life away from society's glare. He settled in a small town and devoted himself to a carpentry business. Some years later, a reporter from a prominent magazine sought to write about the man's life and tell what he dubbed a "fascinating story" to his readers. He met with James, but James remained uncooperative, unwilling to tread back into the painful past. The reporter nonetheless went ahead and published the piece, and James fought the magazine in court to prevent its publication.
>
> Was James's privacy violated? Should James have had a

right to prevent publication of the data? Once the story was published, should James have been able to delete it from the public domain?

—Adapted from *Sidis v. FR Publishing Corp* (1940)

In David Lean's *Lawrence of Arabia*, the film's Colonel Lawrence, an amalgamation of fact and fiction, proclaims that "Nothing is written." Having defied conventional wisdom by successfully traversing the hostile Nefud Desert—and retracing his steps to rescue Gasim, a fallen companion—Lawrence boldly challenges the Fates. He refuses to be defined by the past and instead professes to craft his own destiny.

Proclaimed by the high court of the European Union, the *right to be forgotten* provides a means for each of us to craft our own paths in the modern world. Also called the "right to erasure," as in the 2018 EU GDPR, this principle allows individuals to remove their personal information from the web under certain circumstances. From Australia to India to South Africa to Canada, more and more countries are legally enshrining this right. In the United States, however, where freedom of expression and freedom of the press are fiercely valued, legislatures and courts have been loath to recognize it. To better evaluate its merit, let us take a closer look at the specific contours of the right to be forgotten as defined by the EU.

Defining the Right to Be Forgotten

Google Spain v. AEPD and Mario Costeja González (2014) cemented the right to be forgotten as legal precedent. It began when a man in Spain (Costeja) discovered an online article detailing an auction for his foreclosed home almost twenty years prior. The debt had been paid off, and Costeja asked Google Spain to take down the story, claiming it to be irrelevant and moreover staining his reputation. His case traveled all the way to the EU high court, and with his victory, he opened a floodgate for others eager to eliminate unflattering and expired stories.

In its ruling, the European court established various (albeit vague) parameters for evaluating when the right could be applied. Through the trial, three fundamental questions regarding personal data were asked and answered. First, the court answered to whom the ruling for deletions could be applied: data controllers. This was and remains the designation of search engines such as Google, Yahoo, and Bing, in addition to Facebook. Under the 1995 EU Data Protection Directive, as well as the new GDPR that goes into effect in May 2018, data controllers have many obligations to consumers that are regulated by national watchdog organizations. In this case, data controllers are tasked with extra responsibility in the indexing of links to personal information. Perhaps of significance, news organizations in themselves are explicitly not beholden to a right to be forgotten; nevertheless, if the primary tool for online browsing does not reveal particular articles, the effect is arguably the same.

Secondly, the courts asked whether EU data privacy laws applied to Google Spain, a subsidiary of Google, which has its data processing server in the United States. The court answered in the affirmative, asserting that any company with a branch in a member state of the European Union would have to comply if promoting the "selling of advertising space offered by the search engine." An interesting result of this wording is that a corporation without ad space, such as Wikipedia, need not be subject to the right to be forgotten.[155] Finally, the courts established the criteria for deletion of information: data judged "inaccurate, inadequate, irrelevant, or excessive for the purposes of the data processing" could be successfully challenged and ordered excised.[156] Admittedly vague, these categories are still being defined as increasing numbers of requests go through and Europe's courts continue to shape the right's scope.

In response to the initial ruling, critics quickly protested the potential for censorship. While the court acknowledged that the right

---

[155] It should be noted that Wikipedia would not itself be classified as a data controller, given its encyclopedic function. Hence, it would not have to suppress pages in the first place. However, just as with news agencies, while Wikipedia may not have to suppress pages, search engines still would. In fact, at least fifty Wikipedia links were removed from European browsers in late 2014, according to Wauters, "Right to Be Found."

[156] European Commision, "Factsheet on the 'Right to be Forgotten' Ruling (C-131/12)."

was not absolute and had to be balanced against other fundamental rights, including freedom of expression and freedom of the press, no practicable ledger to balance these different rights was offered.

To decode his concern with the right to be forgotten, American legal scholar Jeffrey Rosen has noted that three types of cases are often conflated in the discourse. These are (a) where someone seeks to delete self-published information, (b) where someone copies or reposts information published by the original subject, and (c) where others post information about a subject. While honoring deletion of the first type is common practice for many tech companies, excision in the latter two is more controversial, and undeniably clashes with others' free expression.[157]

Yet, with the pervasiveness of today's media applications including Instagram, Snapchat, Facebook, Twitter, and various Google products, audiovisual and textual content concerning others can be rapidly disseminated. Posts can be easily posted and reposted. Given the climate of instant sharing, Rosen articulated the fear that requests to delete others' content will also be allowed by the European high court. In turn, this could erode democracy's core by enabling censorship.

One way to evaluate the legitimacy of deleting others' content through a right to be forgotten is to consider existing laws on the books, particularly those against defamation and Prosser's privacy tort of "public disclosure of embarrassing facts." In the US, these have traditionally been applied based upon two criteria: (a) whether the subject is a private or public figure and (b) whether the information revealed is "newsworthy."[158] Though the media's First Amendment guarantees of free speech and expression have typically triumphed over challenges by privacy torts, on a few occasions, individuals have successfully sued for damages. Hence, there appears to be some room

---

[157] Rosen, J., "The Right to Be Forgotten."

[158] Smolla, *Free Speech in an Open Society*.

for the right to enter American borders.

There are a number of methods for deletion available today, including the extreme possibility of attempting to go completely "off the grid." Yet, as the previous chapters indicated, the data ecosystem makes it extremely difficult—though not entirely impossible—to obfuscate one's footprints and evade the gaze of data brokers and others. In the perspectives that follow, we will focus on corporations' possible obligation to remove personal information about individuals. With special focus on American adoption of the right to be forgotten, the perspectives will explore the societal implications of the right. Should any version of the right be adopted in the US of A?

A Case Against the Right to be Forgotten

The right to be forgotten opens numerous risks. One relatively minor concern is implementation of the right. In its transparency reports, Google reveals some of the gross statistics for removals based on government requests. Additionally, it displays sample removal requests (generalized, so as to protect the source's identity) that have been granted and others that were denied. This provides some visibility into the process, but an enormous responsibility is still thrust upon it and other tech companies. Google and others in its position are tasked to serve as judge and executioner for cases.

As a result, abuse by data controllers is one salient possibility: Google, for instance, might take down competitors' content in the near future through the pretext of a right to be forgotten. Alternatively, the size of the regulatory task is logistically crippling. Thus far, Google has received more than five hundred thousand right to be forgotten requests for more than two million websites.[159] Although some guidelines exist, such as the Costeja ruling, much of the legal terrain is still being sketched. Hence, for each of the requests, data controllers' employees must expend

[159] Google, "Search Removals Under European Privacy Law."

time and effort to interpret requests. Compounding the resource problem, many of those forced to lead the analytic process may not have much experience or serious legal knowledge. It would not be unreasonable for more browsers to go the protest route of Yahoo Argentina[160] and remove users' ability to search at all for flagged words.

Since a new infrastructure can be implemented in the United States, modeled on watchdog organizations such as the UK's ICO,[161] this concern can perhaps be met, and in turn, clearer standards enacted. Of course, this shift would still come at significant financial and time costs to companies. Deeper concerns persist, however. For one, the species of censorship promoted by the right to be forgotten stands atop a very slippery slope. Moreover, the misguided overvaluation of digital redemption motivating the so-called right, and the "right's" opposition to Blockchain, further signal that it should ultimately be discarded.

Censorship: A Slippery Slope

Censorship is a slippery slope, and examples of its abyss abound in both the past and present. These cases warn us of how society can be corrupted from even benevolently intended restrictions and how that corruption can thoroughly permeate. We must not begin the terrible tumble.

Historically, restrictions on expressions and thought have been costly. McCarthyism, fueled by a fear that Communism could usurp America from within, threatened to do the very same. In the wake of the Second World War, Communists were blacklisted from the film industry

---

[160] As discussed in Sreeharsha, "Google and Yahoo Win Appeal in Argentine Case," in 2012, model Yesica Toscanini requested that Yahoo Argentina remove photos that she had posted online, but had since spread. Unable to find every last crumb, Yahoo decided to de-list any links with Toscanini's name. Hence, for a time, searching the model's name produced no results; her social identity was deleted.

[161] For more on the Information Commissioner's Office, see the "Further Exploration" section of this chapter.

as well as other areas of society. Any sympathetic utterance toward Communism could result in the loss of one's job and persistent refusals of re-employment. A variety of ringleaders championed this hard-nosed stance, but its fundamental insanity was revealed when Senator McCarthy charged that the US Army itself was "soft" on Communism. A heated confrontation with well-respected army generals revealed how a crazed individual could subvert power and censor others for mere personal gain.

Propaganda has also been very successful at fueling passions with biased data and stoking wars. Prior to and during World War II, the Nazi Party of Germany flooded its citizens with advertisements and feature films that proclaimed Aryan supremacy. The propaganda, shaped by the powers that were, portrayed the Jewish people as selfish, violently anti-Christian, and morally depraved. Meanwhile, across the Atlantic, American propaganda portrayed the Germans as animalistic and rapacious, and the Japanese as shifty and cruel. There were many reasons for the war, of course, but these biased portrayals undoubtedly contributed. Biased portrayals, a serious danger of censored media, can have a severe and violent societal effect.

Today, censorship persists in various nations, and it remains an awful burden. In China, information critical of the state is blocked and expunged. Unpleasant events, including the Tiananmen Square Massacre, are eradicated from the history books. And in 2010, Google discovered a state-sponsored cyberattack intended to infiltrate the Gmail accounts of human rights activists.[162] As a result of such attacks as well as plummeting market shares due to homegrown competition potentially sponsored by theft and encouraged by the Chinese government,[163] Google has effectively been forced to pull out of China. Though the company has plans to reenter the massive Chinese market, the fact of

162 Waddell, "Why Google Quit China."

163 Culpan, "Google Won't Play in China."

censorship remains; according to Chinese government dictates, Google's reentry would be contingent upon compliance with censorship rules.

Similarly, in Turkey, libel and defamation suits are stronger than in the United States and even Western Europe. If public officials are criticized, they can successfully sue on minimal evidence, buttressing their power and control of the state. The dominant news media and textbooks deny that the Armenian Genocide ever took place, while journalists reporting otherwise can be incarcerated or executed. Likewise, censorship challenges abound in Iran (where Twitter has departed) and Russia, not to mention the former USSR and various Communist governments of the past.

The oppressiveness of these regimes warns us not to slip down the path of censorship. Even if momentarily convenient, even if apparently necessary, the risks amplify over time. As power is granted to shape conversations and discourse toward the desirable rather than the true, that power can be siphoned by a single corrupt individual and never relinquished. Truth can be locked out so that two and two make five.

Some might say that this criticism is extreme: some forms of censorship are permitted and required by the law. After all, though the First Amendment grants freedom of expression in the United States, falsely yelling "fire!" in a crowded movie theater is still famously prohibited.

There is a fundamental difference here, however: a difference in kind. Prohibitions on speech activate because of the "public harm" they inflict. The harm in yelling "Fire!" in a crowded space is measurable: a stampede and quantifiable casualties will result. Negative effects also include emotional harm via hate speech, but on the whole, as most such cases have been rightfully reasoned in American courts, offensive speech should not be legally prohibited.[164]

---

[164] See the "Further Exploration" section for more on the nuances of "hate speech" legislation.

SPOTLIGHT: DOES ACCESS OF THE INTERNET ALWAYS PRECLUDE CENSORSHIP?

For corporations that value free markets and free expression, censoring nations raise an important concern—is it better to offer some access to information while complying with censorship in the hopes of improvement over time? Members of the ACLU have championed the affirmative on this question (Waddell, "Why Google Quit China"), but it comes with significant risks to a company's image and potentially its services as well.

The global intent of right to be forgotten proponents further indicates that this imagined right must be nipped in the bud. Already, the right to be forgotten has spread virally and evolved into something stronger. Whereas the initial purview of the right to be forgotten in Europe involved priority indexing of pages, circa 2010, the right codified in the GDPR effectively relates to access of content itself. Search engines are tasked with removing offending pages entirely from their results. Other sites that link to the offending content must also be deleted. Though still being sketched in courts, the right to be forgotten as presently conceived extends all the way into Rosen's third sphere.

Moreover, given the global nature of the Internet, changes in one location affect the content presented across the world. Following the Costeja ruling, Google agreed to not display offending pages to anyone using an affected country's server—for instance, Costeja would apply for Google Spain as well as for the branches in all other EU member states. Quickly, this expanded to preclude any individual in Spain (or the EU) from being able to access the offending pages, even if using the US Google website and server. Since then, the right has morphed to potentially encompass expulsion on all of Google's servers regardless of a user's location.

In an ongoing court case still being heard in the French high court as of the summer of 2017, proponents of the right to be forgotten seek global dominion. If these proponents succeed, Google would have to expunge links not only on local servers but across all their servers. According to France's privacy watchdog organization CNIL, the right to be forgotten will have negligible impact until it is globally enforced. CNIL argues that the intent of law is to protect the consumer, and if the irrelevant information remains at large, then significant negative repercussions can still emerge: a cousin in Portugal or prospective in-law living in America, for instance, will still have access to the harmful information.[165]

The counter to this line of reasoning is fairly obvious and one that Google has passionately articulated. Besides the inherent danger of censorship, there is also the precedent. If a nation can determine what news is fit to print, then extremely restrictive regimes of countries such as China and Turkey can further limit what others in the world can access. One solution would be for Google to simply reject such restrictions and leave offended countries. However, in that case, the fruits of the Internet would be completely closed off to large populations, arguably an unreasonably steep price to pay.

A number of questions remain about implementation of the right to be forgotten, and as recently as February of 2017, France's high court has sought greater clarification on criteria for "forgetting" from Europe's high court.[166] These criteria will influence how the court rules in this case. For the sake of free speech, we can hope that global censorship fails to procreate. Otherwise, the slippery slope will be primed to go.

---

[165] CNIL, France's data privacy regulator, astutely noted that selective deletion of information online would limit the right to be forgotten since family or friends located abroad might still be able to access webpages deleted in any particular country (see the French infographic at CNIL, "Portée du Déréférencement de M. Plaignant Appliqué par Google"). CNIL's dissatisfaction with this state of affairs has inspired its push for a right to be forgotten that is globally applied. This case, lodged on August 21, 2017 remains to be settled with the latest documents at http://eur-lex.europa.eu/legal-content/EN/TXT/?uri=CELEX:62017CN0507.

[166] Conseil D'État, "Right to Be Delisted."

Digital Redemption is Overvalued

With the pervasiveness of social media, users can post instantaneously, and data is difficult to delete entirely. Hence, one of the chief motivations for implementing a right to be forgotten is that individuals should not be unfairly judged on prior mistakes. Nevertheless, though this sentiment is noble, the mechanism is inadequate for many reasons.

For one, a world where past misfortunes are commonplace equalizes everyone. Revealing financial and social struggles would devalue their significance. If anyone's sordid history can be accessed, no one has leverage over others. Instead, this openness of information allows for society to better understand its constituents. Everyone errs, and everyone has regrets. Those regrets need not be hidden out of shame, and through revelation, redemption and relief are possible.

Moreover, excising a history does not excise the truth of it. The past offers lessons to be learned, and hence, progress can be made. A society that can choose its own reality by hiding its dark side threatens to delude itself dangerously and perhaps explode violently in the future.

Fundamentally, the problem with judgment does not lie in the access of information. Rather, it resides in how that information is used. Insurers incorporating data about a citizen's foreclosure long ago, for instance, are perhaps abusing the data. In incorporating the data without consideration of character development or mitigating circumstances at the time, the evaluations are incomplete. The insurer reduces a rich story to a single negative data point and potentially loses other important data in the process. Making the ledger even more incomplete by removing the past via a right to be forgotten is not a solution. Instead, a greater focus should be placed on transparency of important processes for consumers and better incorporating holistic data points.

Emotional Harm Must be Evaluated in Context

When it comes to censorship of hate speech, there are two main justifications. First, there is fear that exposure of the hateful text to the public will result in its transformation into violent action. However, in our democracy, political power ultimately flows from the people, and to subscribe to this fear would be mistaken. It insults and fails to comprehend the power of the rational mind.

The second fear is that the magnitude of emotional harm caused is greater than any other concern. In many cases, this has served as the justification for applying the right to be forgotten. Certainly, emotional harm is nothing to scoff at and can be very real—experts estimate that through cyberbullying, the rate of suicides has recently increased by 30 percent in the United States.[167] Yet, the long view needs to be considered. By banning or deleting a work and thus preventing people from making up their own minds about the offensiveness of certain content, a government prevents the public from exercising its reasoning capabilities. A truly offensive work should be identifiable as offensive based on its content—failing to respect human dignity, for instance, or advocating senseless and myopic violence. Shielding the public from such an offensive work does a disservice to that population. Instead, through exposure to an offensive idea, its vileness becomes more pronounced and describable. In the end, grappling with the opposition, however vile, will lead to a deeper understanding of the underlying motives (e.g., shame, fear, or impotency) and moreover help to combat universal ills (e.g., myopia, xenophobia, and faithlessness). Truly grappling with these offensive ideas allows individuals to exercise their autonomy. It empowers them and those affected by the hateful speech to prevent vile policies from implementation. In the long run, exposure benefits society. Truth and transparency are of the highest value.

Certain offensive and exploitative measures are currently prohibited in the United States and across the globe. Ads for payday loans, for instance, are prohibited on Google because of their exploitative tendencies, as discussed in Chapter 3. Moreover, violations of copyright law are taken down by Google and punished where possible

[167] Jones, *Ctrl + Z The Right to Be Forgotten*, 5.

on any Google domain. Where restrictions are not patently necessary, however, the value of free speech is rightfully enshrined first and foremost in the US Bill of Rights. A free press must be maintained. For, the unfamiliar is also the seed of innovation. America was founded on a revolutionary idea, and revolutionary ideas are the key to progress. The right to be forgotten threatens to institute a global and pervasive censorship. It threatens innovation. It threatens true progress.

Building the Blockchain

Besides the many risks of censorship raised by the right to be forgotten, it also directly threatens a valuable technology: Blockchain. Chapter 8 will explore in more depth the very serious danger of centralized databases. Suffice to say here that for a variety of reasons—abuses by data controllers gone rogue, spirited hackers, and widespread malfunction—merely having centralized databases with citizen information is dangerous.

One significant alternative to such databases involves use of Blockchain, a technology that many investors currently see as the way of the future. Experts estimate, for instance, that more than $1.4 billion was invested in 2016,[168] only eight years after its conception.

An intriguing technology for its novelty, Blockchain also holds great promise. In Blockchain, data is not stored locally. Instead, it is spread broadly and compounded. Hence, Blockchain potentially offers increased security via decentralized data storage.[169] Furthermore, one popular use of Blockchain is Bitcoin, which promises a faster means of transaction with quicker authentication and arguably increased security. As another measure of the technology's growth, a single coin that was

[168] Kennedy, J., "$1.4bn Investment in Blockchain Start-Ups in Last 9 Months, Says PwC Expert."

[169] For a deeper introduction to Blockchain, see the "Further Exploration" section of this chapter.

worth nothing eight years ago is now worth $2,600.[170] Hence, besides increased security via decentralized storage, Blockchain also promises benefits including faster processing time and easier online transactions.[171]

Blockchain technologies necessarily save information at nodes and multiply it out, meaning that data can't be easily deleted. As such, Blockchain (at least in its current conception) stands opposed to the professed right to be forgotten.[172] Of course, technology should not dictate policy, but the benefits of Blockchain are yet another argument to tip the scales in favor of eliminating the right to be forgotten.

A Case For the Right to be Forgotten

In 2014, 61 percent of Americans registered support for some form of the right to be forgotten. This included 21 percent of all respondents, who supported special application only for minors or else nonpublic figures.[173] In 2015, the State of California adopted a "right-to-be-forgotten" law for citizens under the age of eighteen. Though the Californian law has a smaller purview than in Europe, applicable only to content directly posted by the minor,[174] it signals that US citizens are serious about enshrining their own right to be forgotten. The wide spread of legislation related to the right to be forgotten signals that this

---

[170] Blockchain, "Market Price (USD)."

[171] Farmer, "Blockchain Technologies and the EU 'Right to Be Forgotten'—An Insurmountable Tension?"

[172] Lumb, "Downside of Bitcoin: A Ledger That Can't Be Corrected."

[173] Humphries, "US Attitudes Toward the 'Right to Be Forgotten.'"

[174] For a lengthier discussion, see Groden, "What Is the 'Right to Be Forgotten' and Will It Ever Exist in the United States?" and the "Further Exploration" section of this chapter.

sentiment is part of a global trend. Majority opinion may not be sufficient in itself to justify widespread right-to-be-forgotten laws. However, understanding more deeply the underlying reasons for the majority's opinion only strengthens the case for the right.

While the EU currently only requires subjugated corporations to excise their page indices (leaving data or even links to pages intact), it is clear that the ideal is broader than that. Nonetheless, this is not as devastating as critics suggest it to be. Rather, it reshuffles the priorities of the Internet in a way that empowers and benefits citizens. Ultimately, for a number of reasons including the inadequate protections offered in the status quo, the true satisfaction of rehabilitation's promise, and the need to return power to the people, the right to be forgotten is essential. Moreover, the value of Blockchain can still be compatible with the right.

Harm Must be Prevented

In the United States, a form of the right to expungement explicitly exists for official records. The erasure statute in many states allows for the state to strike from the record and destroy information about dropped lawsuits. Lorraine Martin of Connecticut faced such a scenario when the state officially dropped her arrest, and she can legally swear in court that she was not arrested.

Yet, Lorraine's situation also highlights the inadequacy of protections in today's digital world. While the official record was changed, news articles about her arrest containing details of her case remained available online. Lorraine has plausibly claimed that the availability of the information hindered her from finding employment for at least two years after the vacated arrest.[175] Nonetheless, her 2015 suit against the news companies on grounds of libel, invasion of privacy, and infliction of emotional harm was quickly dismissed by the justice system, just like many others of the same kind.

Quantifying specific uses of the Internet is undoubtedly

[175] Frankel, "No 'Right to Be Forgotten' Even if Record Is Expunged: 2nd Circuit."

imperfect, but estimated figures that 80 percent of employers and 30 percent of universities search online for additional applicant information[176] seem reasonable, if not conservative. Ultimately, information available via the web can impact one's financial circumstances and future opportunities: the content available is extremely significant. Though censorship is an important concern, the slippery slope only extends as far as citizens will allow. Thus far, consistently, public citizens are prevented from carefree expungement, even in Europe, and we can expect that to remain the case: in fact, of the nearly two million URLs Google has evaluated since the Costeja case, only about 50 percent have been removed, indicating that checks remain. Moreover, as of 2015, fewer than 5 percent of requests have come from public figures,[177] signaling that overall, the law has not been misappropriated to eliminate relevant information.

The current practice of having search engines serve as judges and executioners is undoubtedly inadequate. However, this is a problem of infrastructure, not of the law's design. The infrastructure of evaluating requests will need to be enhanced as requests increase, allowing both greater guidance and oversight.

With the emotional and practical impact of still-circulating irrelevant information, establishing a right to be forgotten is important. This is further highlighted by one of the original motivations for establishing the principle: citizen empowerment.

The Playing Field Must be Equalized

Legality and morality have a complicated relationship, and the precise societal role that the law plays continues to be vigorously debated. It may serve to closely represent a people's morality or else serve a different function altogether, as a check on selfishness and

[176] Jones, *Ctrl + Z The Right to Be Forgotten,* 5.

[177] Tippmann and Powles, "Google Accidentally Reveals Data on 'Right to Be Forgotten' Requests."

depravity or a means to champion particular principles. While there may be numerous universal principles with moral force, such as prohibition of murder, there is also bound to be much variation in the fringe cases, which is where a majority of the right-to-be-forgotten cases tend to play out. For instance, two people may well disagree on the newsworthiness of a story about William J. Sidis.

SPOTLIGHT: THE TRAGIC COSTS OF CYBERBULLYING

Tragedy has sprung from cyberbullying, and in several high-profile incidents, images broadcast and spread on the Internet have been the source of pain. Tyler Clementi's is perhaps one of the most tragic. Eighteen-year-old Clementi was a student at Rutgers in 2010 when his roommate recorded a sexual tryst without his consent. Boasting of the footage via Twitter, the roommate posted it online. Two days later, a shamed Tyler committed suicide off the George Washington Bridge.

More on Tyler's story as well as ways to combat cyberbullying can be found at Tyler Clementi Foundation, "Tyler's Story." Though bullying may have old roots, heartbreaking cases like Tyler's, Amanda Todd's, and many others, are made possible by today's streaming technology and the effectively permanent nature of information posted online. Given such tragic consequences, how should we value the right to be forgotten, and can fears of a slippery slope be outweighed by these common tragedies?

The idea that an utterly transparent world with all transgressions revealed will be a just one is also painfully naïve. The rich and powerful will always have extra means to avoid association with the hordes. Presidents FDR and JFK, for instance, were able to keep their extramarital transgressions secret, even as "yellow journalism" revealed intimate details of movie stars and fashionable contemporaries. In turn, the right to be forgotten actually offers a remedy: it allows individuals to defend themselves with dignity, just as the most well-connected and well-loved do.

The world of utter exposure is furthermore not a valuable ideal. Inherent to any rational creature are thoughts that stray, at least at times, from the norm. In a world with secrets exposed, besides utter honesty, there might alternatively be utter deceit. Like Winston Smith of George Orwell's *1984*, we may have to train our eyes to never betray what lies within or speak aloud what we truly feel in order to "progress" toward that ideal citizen. Exposure of exposure might ultimately lead to a situation just as distressing. Beyond this potentiality, there also appears to be something of the mischievous ingrained within us. And with our discriminating tastes, we harbor minor dislikes and likes that we may not express. What would be gained from revelation of these dislikes? If Eliot the friendly acquaintance discovers my slight dislike for his nasally vowels, does that truly serve anyone? Rather, it foments grudges and imparts wounds that need never arise at all.

Finally, it is also necessary to note that the laws do not grant unchecked excision of data. There is a process, and however vague the initial Costeja ruling was, the principles are unlikely to apply to public figures so that the data obviously relevant to the public—the imprisonment of a government dignitary for scandal, for instance—will not be forgotten. The most important transgressions will remain publicly accessible. As a result, the citizen and society will be empowered as never before.

Enabling True Rehabilitation

The original motivation of right-to-be-forgotten legislation can be traced to a 1970s-era UK measure to grant rehabilitated convicts a true chance to commence again. The Rehabilitation of Offenders Act (1974) in the UK was intended precisely to prevent past histories of rehabilitated criminals from affecting future insurance or employment, and this sentiment is ever more relevant now.

With the Internet and social media more prolific than ever, information is currently stored and almost instantaneously disseminated. How often has it been said that indiscretions of youth should not have a bearing upon the future? Certainly, as formulated, the right to be forgotten also has a wider berth than mere youth, as it concerns even recent errors; nevertheless, the important sentiment remains: the truth of rehabilitation ought to be realized.

At the core is a belief that humans are redeemable. Do we not applaud the sinful heroes of film and literature who deeply sin but eventually repent? Do we not esteem them for their courage and journey? *Casablanca*'s Rick finds meaning in finally acting selflessly, as does knight Antonius Block of *The Seventh Seal*, Amir in *The Kite Runner*, and Mary Magdalene. In Immanuel Kant's formulations of the categorical imperative, he discussed at length the moral beauty of those without proclivities to moral action who nonetheless abide by it.[178] To sin and err is human and natural, as many cultures profess. Perfection is unattainable. Using our rational capacity to right the ship and move forward is perhaps among the most meaningful acts we can take.

This idea of redemption is not appealing simply for some aesthetic beauty; it also plays an essential part in modern legal systems. The death penalty has increasingly been abolished, chiefly because humans ought to be able to regret and atone for sins. And neither is this sustained living simply to avoid the stain of blood upon one's hands—instead, full rehabilitation in many cases remains the goal for the benefit of the individual as well as society.

Given the dissemination possible through the Internet, employers and insurance agents can discriminate with the wealth of information available as never before. Company protocols may nominally insist that information found on the web does not influence such distinctions, but how can that actually be enforced? If excising the data does not harm anyone and bears minimal public harm, then should not the person exposed be granted control?

[178] Kant, *Grounding for the Metaphysics of Morals.*

Blockchain's Value and Alternatives

Regarding Blockchain and its vast potential, technology should not dictate the law; power flows in the opposite direction. Certainly, technology can offer new benefits and potentials, but the science must progress in a morally and legally sensible way.

While the benefits of Blockchain have been fairly stated, especially the decentralization of information—a major, potential privacy boon—it is also worth noting that Blockchain does not absolutely preclude the deletion of subsequent information. All expanded nodes will have to assent to modification of base information, but if some mechanism can be established, this need not be impossible. Of course, a mechanism to modify information after entry into Blockchain fundamentally impacts the current conception of the service; nonetheless, as a nascent technology, there is room to still shape it to best fit our holistic needs.

This last point is worth reiterating: at present, the ideal implementation of Blockchain remains a dream. A limited version of it is starting to develop and undoubtedly will continue to do so. However, it will not materialize for some time, and even then, it is unlikely to emerge with all the desired properties ascribed by enthusiasts.

Additionally, Blockchain would not be an all-or-nothing affair. It could be used selectively to deal with particular protocols without compiling personal data. Realizing this will be a challenge, especially given the widespread uses that its exponents have in mind; again, however, given the nascent stage of the technology, we should pay attention to what we want to be possible and appropriately evaluate the risks. We should develop the technology most amenable to our desires and not be driven by an eager minority that has not performed such an adequate evaluation.

In Conclusion

In the EU's GDPR and in the formal legislation of other nations, the right to be forgotten has been codified. Nonetheless, if the right seems draconian, the spread can still be checked in the United States and the formulation reconceived. The difficulty for global corporations that seek customers in jurisdictions where the right currently or soon will apply is that they must have a mechanism by which to abide by the rule of law or else risk a fine an the people's ire. Google and other data controllers have begun to abide by this requirement, albeit with some growing pains; how widespread these capabilities can and will become remains to be shaped by those most deeply affected by the right: us.

William Sidis lost his case against the newspapers in 1940, and four years later, his life. Yet, the war in which he engaged still rages on. Should a private individual have control of disseminated information about himself? Or, should the power of the people to independently decide what is important and newsworthy prevail? How the United States rules will signal to the world where it stands today and how it wishes to shape its future. Moreover, the impact of cases to be ruled on globally in France, India, Australia, and Africa will further shape the future for decades and perhaps centuries to come. The year 1984 came and went without Orwellian fears of big government being realized; meanwhile, countless trials of embarrassed individuals have persisted. Yet, the risks of surveillance states are always significant, and the next chapter will explore these in more detail.

Further Exploration

As discussed in this chapter, *Google Spain v. Costeja* (2014) has been instrumental in shaping the right to be forgotten. The full and **final Costeja ruling** can be found at **http://curia.europa.eu/juris/document/document_print.jsf?doclang=EN&text&pageIndex=0&part=1&mode=DOC&docid=152065&occ=first&dir&cid=437838**.

The **latest EU press release** highlighting Costeja's importance can be found at https://curia.europa.eu/jcms/upload/docs/application/pdf/2014-05/cp140070en.pdf, and a **short summary** sponsored by the EU is available at **http://ec.europa.eu/justice/data-protection/files/factsheets/factsheet_data_protection_en.pdf**.

In the ongoing case ***Google Inc. v. CNIL*** (C-507/17), the EU's high court will rule on whether the European "right to be forgotten" requires companies to modify searches in non-EU countries as well. The latest case documents can be found at **http://eur-lex.europa.eu/legal-content/EN/TXT/?uri=CELEX:62017CN0507**.

The full text of the **2013 California bill granting children a right to be forgotten** can be found at **http://leginfo.legislature.ca.gov/faces/billNavClient.xhtml?bill_id=201320140SB568**.

As required by the European Union, each member state has its own independent organization to investigate potential breaches and uphold data privacy for the country. These are "data protection authorities" (DPAs). The **Information Commissioner's Office** is the UK's organization, and its site features articles, news, and actionable updates about data use in the UK and across the globe: **https://ico.org.uk/**. The equivalent organization in France is **CNIL** (La Commission Nationale de L'Informatique et des Libertés): **https://www.cnil.fr/en/home**.

Blockchain is a complex technology, and much more remains to be said, while still more has yet to even be discovered. One introduction to the technology, along with a more holistic discussion of benefits and risks, is **Dan Tapscott's interview with McKinsey's Rik Kirkland** (http://www.mckinsey.com/industries/high-tech/our-insights/how-blockchains-could-change-the-world). Tapscott's book, ***Blockchain Revolution: How the Technology behind Bitcoin is Changing Money, Business, and the World***, coauthored with his son Alex, goes into even more depth.

Hate speech in friction with the First Amendment has been the source of many heated historical battles and much oracular argumentation. *Hustler Magazine, Inc. v. Falwell* (1988) is one of the most recent, explored in Rodney A. Smolla's eminently readable account ***Jerry Falwell v. Larry Flynt***. This case involved a lawsuit by televangelist Jerry Falwell against a satirical and highly off-color ad in the adult magazine *Hustler*. After a lengthy journey featuring numerous sharp twists, the case reached the US Supreme Court. There, the high court unanimously ruled in favor of *Hustler*, though not without articulating various conditions that perhaps set free-speech boundaries.

Chapter 6: Drones, Government Surveillance, and Citizen Privacy

Key Questions:

- What is the extent of government surveillance, including by domestic drones, today?
- What data can be collected by governments? How much does this data reveal?
- Who, if anyone, should be allowed to surveil citizens—and under what circumstances?
- What are some of the key profits promised by drones? What are key dangers?
- What are the benefits of surveillance? What are the major costs?
- What protections can ensure drone technology is deployed appropriately?

> Consider: A hummingbird drone deployed by the National Security Agency buzzes through a boisterous US city. Through audiovisual sensors, the drone monitors for violent civil unrest. It analyzes facial expressions, body language, trigger words, and tone of voice to predict violent riots and terrorist threats before they erupt. Constantly compiling data, this drone and others of its charm are always monitored, though typically not by humans; instead, their findings are examined through automated AI analysis back at headquarters and local stations. In the event of potentially troubling developments—and only then—human operators are notified, enabling them to take swift preventive measures, beginning with contact of local law enforcement.

If you were aware of its mechanical innards, what would the sight of a hummingbird drone evoke in you? Joy, for the increased security the machines provide? Indifference at something unlikely to impact your daily life? Fear at the prospect of discovery—the unearthing of your private activities, even if not illicit? Some combination of these? Or, something else entirely? On the whole, does such surveillance appeal to you?

—Inspired by current drone technology and ongoing techniques to improve sentiment analysis

Articulated by Brandeis and Warren as the right "to be let alone," privacy has been applied to a variety of subjects, including corporations and individual citizens, and to a variety of concerns ranging from the economic to the religious and the intellectual. It has also been applied to both the physical and mental realms. Yet, of the many descriptions bandied about, *privacy* is perhaps most intuitively understood in terms of private physical behaviors. It is here that drones have recently swarmed in with their unprecedented surveillance capabilities. As with any technology, there are competing dangers and benefits, and this chapter seeks to lay bare the key tensions.

Defining Drones

More formally known as unmanned aerial vehicles or UAVs,[179] drones are pilotless aircrafts that are remotely controllable or otherwise fully autonomous. They may possess weaponized payloads and often have video and sometimes audio capabilities. Although the predominant critical focus has been on manually controlled drones, the recent feasibility[180] of self-operating machines means that these will also need to be considered in future drone policy.

Drones are perhaps most familiar from their use in military strikes abroad. Their increased use in assault has provoked a fiery debate. On one hand, military drones reduce risk to their users' armed forces (typically America and her allies today), and potentially provide a more comprehensive response than ground personnel could likely offer. On the other hand, missile strikes have an inflexible target radius, and as such, military drones' mechanical indifference means an increased risk of civilian casualties and other collateral damage to innocents. Some critics also question the accuracy of the rapid identification technologies involved, as well as the psychological impact of continued faceless

---

[179] Sayler, "A World of Proliferated Drones: A Technology Primer."

[180] Bernstein, "Top 5 Best Auto-Follow Drones."

combat. While military drone operators have a systematic kill chain of command—including repeated corroborations before strike activation—this does not preclude judgment errors, especially in the heat of the moment. Yet, given the stakes involved, even a single mistake may be unacceptable.[181]

A complete discussion on military rules of engagement may be particular to wartime, but the core concern for national security informs homeland drone use as well. Domestically, non-weaponized drones are rapidly expanding for both commercial and government purposes. The Federal Aviation Administration (FAA), governing body for American aviation, has estimated that by 2030, thirty thousand UAVs will regularly fly in the nation's skies.[182] As even drone champions have acknowledged, such proliferation poses a threat to citizen privacy. Through widespread surveillance, public spaces become even less private, while private ones can be intruded upon more easily. Fears along these lines have undoubtedly motivated the restrictive drone laws present in more than twenty-six states.[183] Yet, as UAV champions have noted, drone technologies are also not currently as invasive, nor is the data they collect as informative, as what private corporations gather via smartphones and web browsing habits (see Chapters 2 and 3). Instead, advocates point to the myriad benefits of UAVs, including bolstered national security as well as likely economic and knowledge boosts. Hence, as with all technologies, there is a fine balance to be achieved.

Part of this balance is technical in nature—for all drones, computing power is fundamental.[184] Advancements in computing power have thus far easily exceeded the predictions of Moore's Law, which

---

[181] For a compilation of prominent arguments for and against the use of military drone strikes, see the Further Exploration section of this chapter.

[182] Thompson II, "Drones in Domestic Surveillance Operations."

[183] Risen, "FAA Considering Drone Flights Over Crowds."

[184] Springer, *Military Robots and Drones: A Reference Handbook*, 1-29.

states that processing power doubles every two years.[185] Commensurately, capabilities are rapidly improving. Computing power operates the machine, but to function adequately, a drone must also be well-engineered. This engineering, too, has improved substantially over the years. For instance, in 2005, the electro-optic sensors on some drones could "identify an object the size of a milk carton from an altitude of 60,000 feet."[186] While weather as well as geographical conditions can affect the images collected, improved cameras have improved resolution. For instance, in June 2016, EPIC noted that "with special capabilities and enhanced equipment, drones are able to conduct far more detailed surveillance, obtaining high resolution picture and video, peering inside high-level windows and through solid barriers, such as fences, trees, and even walls."[187] Chiefly, one area of focus for drone usage has been improving collision safety, similar to how engineers are seeking to improve the capabilities on autonomous cars, and as indicated in its latest report, the FAA is satisfied enough with progress to champion wider use of UAVs in public spaces.[188]

Another aspect of this balance is legal. Drones, though in some sense miniature airplanes, are a novel technology without precedent. A few cases offer piecemeal guidance, notably 1942's *United States v. Causby*, generally seen as the birth of aviation law. In *Causby*, a North Carolina chicken farmer successfully sued the US government for using low-flying military aircraft that he claimed damaged his livelihood. In a 5-2 verdict,

---

[185] Waldrop, "The chips are down" explains how transistor chip size is beginning to slow, but this is a function of quantum uncertainties that activate below a certain size. Progress may hence be slowing but is still well above the Moore's Law standard. Other elements related to computation power continue to improve, such as speed through greater interconnectivity and energy efficiency.

[186] Electronic Privacy Information Center, "Spotlight on Surveillance."

[187] Hood, "FAA Gives Commercial Drones Takeoff Clearance."

[188] Ibid.

the Supreme Court ruled in favor of the citizen and established that airspace as high as 83 feet above a residence was private property. Since then, the FAA's 2016 guidelines for drones have become the standard-bearers for UAVs. Though states may provide more specific laws curtailing drone use, these guidelines provide the benchmark. Nevertheless, the guidelines' production history has been one of contention, undergirded by fierce challenges.

With drone technology burgeoning in 2012, Congress issued a directive[189] to the FAA to produce "comprehensive" rules for drone usage. Precisely how comprehensive these new rules were supposed to be was only hinted at. While prescribing certain safety considerations to be included, such as that every drone have a "sense-and-avoid" motion capacity, the act did not explicitly detail privacy or many other considerations. Owing to their concern, public-interest advocacy firm EPIC[190] submitted a petition to the FAA featuring over one hundred signees, including individual experts and representatives from organizations such as the American Civil Liberties Union (ACLU). In their petition, the signees called for special attention to be paid to relevant privacy questions.

In June 2016, when the FAA released its completed rules, it confirmed EPIC's fears.[191] The rules provide many important technical safeguards, including the specifically commissioned preventive measures to avoid collisions. Overall, the regulations widely promote increased drone usage. However, while acknowledging receipt of EPIC's privacy concerns, the report indicated such considerations were outside its scope. EPIC has since resubmitted[192] a suit against the FAA, results

[189] U.S. Congress, House of Representatives, "FAA Modernization and Reform Act of 2012 Conference Report."

[190] Electronic Privacy Information Center, "EPIC v. FAA"

[191] Federal Aviation Administration 2016.

[192] Electronic Privacy Information Center, "EPIC v. FAA."

pending as of this publication. For their part, some government leaders have repeatedly indicated a desire[193] to uphold privacy rights, having recognized their significance; how this plays out in practice, however, is a story still unwritten.

The future possibilities of drones include much more than what is sketched above—there are things already envisioned yet prohibited, and there are dreams still unthought. The former includes drone package-delivery services[194] via such popular carriers as UPS and Amazon. Although presently prohibited since pilots must be within the line of sight of aircraft—conceivably an effort by the FAA to meet some privacy concerns—if and when regulations change, autonomous drones will also begin to see widespread use.

Specifying Scope

Particularly after the 2013 Snowden leaks, government surveillance has been a topic of great public contention. While there has always been a vocal minority wary of government intrusion, the disclosures harnessed a keen public focus on the issue. As the disclosures revealed, the agency widely collected cell phone metadata, conducted e-mail scans, and stored collected data in accessible data banks through the cooperation of telecom and other technology companies.[195] Moreover, all of this occurred without the knowledge of Congressional representatives, including those specifically designated to oversee surveillance activities.[196] Suspicion of surveillance additionally extended

---

[193] White House, "Presidential Memorandum"

[194] Von Zeipel, "FAA Issues New Rules for Commercial Drones."

[195] Szoldra, "This is Everything".

[196] See, for instance, Senator Diane Feinstein's change of heart after learning that NSA surveillance extended to the president, members of Congress, other world leaders, and other countries' citizens as reported in P. Lewis and Ackerman, "NSA: Dianne Feinstein Breaks Ranks to Oppose US Spying on Allies."

to senior government officials; in September 2016, for instance, former FBI Director James Comey recommended that public users follow his example and cover their laptop cameras to defend against potential snoopers.[197]

Numerous other writers and privacy professionals have described in depth prominent mechanisms of government surveillance and emergent concerns.[198] This chapter narrows the focus to surveillance possibilities via drones. A particular means of surveillance as well as other explorations, drones highlight a chief conflict at the heart of surveillance: law enforcement and public safety interests weighed against citizens' privacy. The case sketched at the chapter's start focuses on security and combines it with the budding practice of sentiment analysis, also known as "opinion mining." More commonly employed by advertisers and corporations like Facebook, this technique uses some platform, typically social media, to analyze the "happiness" of individuals. As reported by the deputy director for digital innovation at the CIA in October 2016, it is also being pursued by the CIA.[199] The discussion below will explore the intertwined issues.

A Case For Rapid Drone Proliferation

For their enormous security potential, and also for their economic and intellectual promise, drones should be propelled still higher. With their unique technical abilities, drones could effectively check the recent and frightening increase of public attacks. Moreover,

[197] Hattem, "FBI Director: Cover Up Your Webcam."

[198] See the "Further Exploration" section of this chapter for more.

[199] At least as far back as 2011, the CIA analyzed social media to understand general sentiments and proactively react to threats (see, e.g., Keller, J., "How The CIA Uses Social Media to Track How People Feel"). Since then, efforts have expanded and increased in accuracy, as reported in late 2016: Konkel, "CIA Can Anticipate Social Unrest 'Three to Five Days' Out."

drones promise to boost other trends, productive and progressive in nature. For instance, they can enhance scientific exploration in cost-effective ways, improve health services, and significantly strengthen economies. At the same time, the prospect of totalitarian futures has been vastly overstated by critics; as has been indicated by instantiations of increased surveillance within democracies, negative effects can be avoided through checks such as oversight and public transparency. By focusing on the massive benefits and instituting some minimal regulations, drones can soar to achieve their immeasurable value—a value for which even the sky will be no limit.

National Security Concerns

Compact and capable of wearing familiar disguises, drones are uniquely powerful devices. Being small and typically agile, they can easily enter crevices and broach blind spots that mounted surveillance cannot. The 2018 FLIR Black Hornet PRS is one such model, a pocket-sized drone weighing less than thirty-three grams, able to withstand twenty-knot winds and temperatures from 14 to 109°F. FLIR is marketing their "combat-proven" UAV for use by field agents in conjunction with covert operatives,[200] which could effectively translate into sensitive missions within the homeland as well. Developers are also focusing on disguised drones that bear resemblance to their nominal animals, and even behave like them. Publicly announced in 2011, the life-size Aerovironment Hummingbird, for instance, has been designed with functional flapping wings and body coloring.[201] Another disguised drone of note is the RoboBee, designed by the Harvard Microrobotics Lab. In development for more than fifteen years, a recent iteration of the RoboBee is able to "fly, dive into water, swim, propel itself back out of

---

200 FLIR, "Black Hornet PRS."

201 See *AirForce Technology,* "Hummingbird Nano Air Vehicle (NAV)," for the official product listing.

water, and safely land."[202] One further machine in development is the WildCat, created by Boston Dynamics. "The world's fastest quadruped robot" can attain speeds of almost twenty miles per hour through a galloping gait that mimics that of speeding dogs and horses.[203] Technically not an aerial vehicle and hence not a UAV, the WildCat is nevertheless autonomous and the underlying technology could be incorporated into flying drones in the near future.

With their mobility, these machines are capable of improved surveillance over traditional grounded models. With their ability to blend into natural surroundings, they are also capable of a more natural surveillance, by which would-be evildoers will be less equipped to shield their nefarious activities. The tremendous value of enhanced surveillance is particularly clear from the events of the 2013 Boston Marathon.

On April 15th that year, residents of Boston and competitors from throughout the world had gathered to witness or participate in a celebration of endurance. Participants would run 26.2 miles after months of training to challenge old records, raise money for humanitarian causes, and celebrate individual fitness. At 2:49 pm, sudden explosions rocked the finish line. Two bursts twelve seconds apart killed three and injured hundreds more.

In the immediate aftermath of the attack, dust and confusion reigned. Blindsided, local and federal law enforcement could not determine how best to begin their investigations. They turned to witness testimonials, but most witnesses, likewise in shock, were unable to recall anything particularly suspicious. From the crime scene, fire forensic experts began working to identify the explosions' source and the materials likely involved. The vast majority of resources, however, was focused on video and photo evidence. Officers canvassed the surveillance camera footage from two hundred nearby businesses.

[202] Burrows, "New RoboBee flies."

[203] Boston Dynamics, "WildCat."

Officers also encouraged citizens who had been in the area to supply their own photos and videos. Ultimately, investigators amassed an enormous ten terabytes of media—almost twice the size of the entire Wikipedia database. Even with hundreds of officers on the case, progress was predictably slow, initially revealing nothing.[204] It was only with the fire forensic results that tangible progress could be made—the results helped to identify the likely type of backpack used by the perpetrators. Focusing on the discovered detail, investigators were able to identify promising photos of the suspected backpacks, albeit with only blurry images of the duo who carried them.

To identify and apprehend the suspects, substantially more remained to be uncovered. Investigators hence decided to crowdsource the investigation, turning the images over to the public, hoping that someone could identify a resemblance. Meanwhile, fearing their discovery imminent, the perpetrators grew palpably nervous, and began a new spree of violence. Passing through the MIT campus, they shot and killed a campus officer in an attempt to steal his weapon. An MIT campus camera captured some of the murder, gathering more precise information on the suspects.[205] Shortly thereafter, the suspects were identified as the Tsarnaev brothers, students enrolled at UMass Dartmouth who had, in their youth, emigrated from Kyrgyzstan to Russia and then the United States.

As the pursuit intensified, one of the brothers was captured, though the other ran him over and killed him. Guessing that the suspect had fled to a nearby harbor, a thermal camera was deployed by helicopter and identified the remaining suspect's heat signature. Armed with the data, police officers closed in, and after tense negotiations for twenty minutes, apprehended the suspect. Altogether, within four days of the

[204] Luna, "FBI Sifting for Clues in Mountains of Evidence."

[205] Feathers, "Middlesex County Prosecutors Building Murder Case against Dzhokhar Tsarnaev."

bombing, surveillance data provided law enforcement a relatively efficient and effective method to capture dangerous suspects-at-large.[206]

Many factors went into ultimately apprehending the bombers, and it was a collective effort from the hard work of hundreds of police officers, emergency personnel, and ordinary citizens. Undeniably, however, surveillance footage was key. Stored visual data provided the first major key to identifying the bombers, with stored images that included backpacks worn. Due to the haystack's volume, the collected data was publicly released to more quickly identify pertinent information; this was valuable in the course of the investigation but may have contributed to the brothers' subsequent murderous activity, alerting them that law enforcement was closing in. MIT campus footage provided further evidence of the brothers' crimes that could be used in trial, and thermal imaging was fundamental to discovering the lone survivor's precise whereabouts.

While this surveillance was valuable, consider some of the ways in which roving drone surveillance could have been superior. During the Marathon, hovering drones could perhaps have served as a psychological deterrent for would-be wrongdoers. Even if the brothers had proceeded, compact drones, operating surreptitiously increase the chance of suspicious activities' discovery. In the bombings' immediate aftermath, stored images from the drones prior to the explosions could have helped to identify the associated parties more easily, and at the very least, identify nearby bystanders whose observations could be queried. Similarly, once the fire forensics returned, law enforcement could perhaps have unleashed drones to search for any identified attributes—provided they were specific enough—even if the suspects' precise identities were unknown. Furthermore, in the final hunt that actually occurred, thermal imaging was conducted by a mounted camera on a

---

206 Kiger,"How They Identified the Bombers: A Timeline From Event to Capture" provides a narrated and somewhat expanded timeline of the events through capture.

helicopter. The camera cost around $300,000, while a commandeered helicopter costs a further $500,000 to $3 million per flight. By contrast, continued drone use cost a federal department just $22,000 in 2016,[207] and new drone models such as the FLIR Black Hornet PRS have particularly keen vision, including electro-optical, infrared, and other sensors. Further sorts of sensory detection can easily be incorporated. Their technical capabilities coupled with compact bodies and unobtrusive design make drones a prime and cost-effective asset for surveillance.

SPOTLIGHT: WHAT IS DRONE FORENSICS?

To deal with increased drone use, law enforcement is developing a new field of inquiry called "drone forensics." In this field, forensics analysts look at a variety of factors, including skin-cell residue on blades and user-account login information that may be stored in the drone's computer (Marks, "How Police Catch Drone-Flying Criminals."). There are limited ways in which drones can store information, but creative analysts can nonetheless uncover a wealth of information. Further regulations by the FCC or FAA may also legally require that certain identification information be easily accessible from a drone computer.

Ideally, these sorts of violent events could be prevented from occurring in the first place. To that end, sentiment analysis affords a potentially valuable tool that could be connected with roving drone operations. In the case of the Tsarnaev Brothers, there was perhaps little to indicate their intentions. On social media, the brothers supported rebels in the Syrian civil war as well as Chechen independence from Russia, though they were far from the only ones doing so. Occasionally,

---

[207] For further specs for the camera used, see Atherton, "How it Works: The Thermal Camera That Found the Boston Bomber." See Kreps, *Drones*, 108 for comparative costs.

they did link to Islamic websites, though not overtly radical ones. In the words of their peers, Dzhokhar and his brother Tamerlan seemed to be "normal" individuals. However, greater focus on the surviving brother's Twitter posts after his trial did reveal periodically angry posts about society and the world as well as repeated jokes about the use of weapons.[208] Perhaps these were regular off-color posts for a young adult, and retroactive vision colors in causality; nevertheless, if there are means of preventing senseless violence, then should they not be pursued?

This especially is the case given the rising number of global attacks. In 2001, the terrible September 11th attacks levelled a communications and economic hub in New York, killing or wounding more than 9000. Frustratingly, several of the terrorists involved with commandeering the United Airlines flights were on national security watchdog lists, but surveillance unto that point had not revealed anything decisive by which to take them down. Of late, suicide attacks have intensified around the globe levelling community centers, churches, mosques, and marketplaces. In Kabul, Tehran, Brussels, London, Paris, Kashmir, St. Petersburg, Melbourne, Maiduguri, Mogadishu, Bogotá, New York, Charlottesville, and far too many other cities, violent attacks have yielded tragedy. From the late 1990s and into the present, the annual number of terrorist incidents has grown from less than one thousand to more than ten thousand. Correspondingly, the number of fatalities has rocketed nearly tenfold.[209]

In part, this devastation may be due to improved technology that makes attacks possible with greater speed and devastation than ever before. Besides explosive devices that might be rigged with spare electronic parts, tiny, even microscopic weapons can also prove fatal. Chemical and biological warfare in particular have been grave fears in

[208] For instance, see Stanek, "The Childhood of Dzhokhar Tsarnaev Was Pretty Normal, Leaving Questions As To What Changed."

[209] See, the "Further Exploration" section of this chapter for the data cited in the last two paragraphs on the impact of airline and other terror attacks over time.

recent years. Anthrax, for instance, seized public attention shortly after the September 11th attacks. Several American reporters received packages with the potent powder (an infectious bacteria) in late 2001, and though no connection was discovered, public fear about bioterrorism reached new highs.[210] More recently, the neurotoxin sarin, which can kill within ten minutes, was used by the Syrian government against protestors during its civil war in 2013; more than seventy died as a result.[211]

Other sorts of attacks, as will be discussed further in Chapters 7 and 8, propagate in cyberspace. These too have increased in power and use, such as with the October 2016 cyberattack on service provider Dyn, the largest of its kind that shut down online service for much of the American Northeast corridor.[212] A few months later, the WannaCry ransomware infiltrated at least 300,000 devices around the world including hospital computers and health monitors in London. The attackers demanded a ransom payment via untraceable Bitcoin, and hundreds paid up, contributing to a haul of more than $150,000.[213] On the whole, incidents of cybersecurity have dramatically risen in number over the last few years.

With so many vectors for attack and the problems yet to be under control, clearly more needs to be done. Technology may in part be responsible for the increased number of threats, but it also offers effective countermeasures. Consider, for instance, the success of the United States Transportation Security Administration (TSA). Created

---

[210] Reaves, "Anthrax: Separating Fear From Fact."

[211] Paton, "What is Sarin Gas, The Chemical Weapon that has Killed Over 70 In Syria?."

[212] Woolf, DDoS attack that disrupted internet was largest of its kind in history, experts say."

[213] See more details on the WannaCry attack at Sherr, "WannaCry ransomware: Everything you need to know."

directly in response to the 9/11 attacks, this agency within the department of Homeland Security was created to oversee airport security. Through additional screening, carry-on restrictions, and consistent airport surveillance the security measures may mildly inconvenience passengers; undoubtedly, however, they have helped to increase public safety. In the years since 2001, the number of airline hijackings across the world have reduced from a regular twenty or more through the 1990s and early noughties, down to the single digits.[214] Clearly, smart surveillance can be highly effective.

Unleashing surveillance drones into the skies, law enforcement will have substantially increased abilities to act rapidly on attacks and minimize harm. They will have a substantially improved ability to serve the community. As enhanced airport security has demonstrated in recent years, and as the Boston Marathon investigation proved, surveillance is essential. Drones provide an enhanced and cost-effective method to surveil public spaces and bolster public safety. Combined with other technologies such as automated analysis sketched out in this chapter's opening, drone surveillance can be even more effective. These benefits come with still another: community knowledge of surveillance will itself serve as a deterrent. Contrary to some critics, this knowledge is unlikely to have other deleterious societal effects.

Surveillance Concerns Are Overstated

In an ideal world, individuals could be left to their own devices. The burdens of bureaucratic oversight could be tossed aside, with budgets streamlined. Governments themselves might be unnecessary. Yet, there is a counterbalance to the "better angels of our nature," one that swims in violence, hatred, and destruction. Throughout human history, individuals and groups do strike, and sometimes that violence stems from within a society. Surveillance hence is something

---

[214] For this and other data, see Roser, Nagdy, and Ritchie, "Terrorism," also linked in the Further Exploration section of this chapter.

fundamental; it may be a tradeoff with some of our civil liberties, but it is a necessary one. Ancient Rome recognized its necessity, with the *cohortes urbanae* who would survey the Forum and other public areas for signs of disorder. Free expression was valued, but violent outpourings in the public forums were also common and guards at the ready could minimize damage before severe escalation.[215] Modern surveillance may use fancier technologies than those in the past, but the intent remains the same: to increase public safety. As one recent example of increased surveillance indicates, this intent can coexist with modern democracy.

A prominent fear regarding increased surveillance is the "Chilling Effect" by which free expression might be stifled. The reasoning goes that citizen knowledge of surveillance would induce fear and stifle legitimate but unpopular activities, conversation, and thought. In turn, this would induce a stagnant state of conformity. However, the hypothetical causal chain does not seem to be supported by actual fact.

Consider London, where public surveillance increased around the turn of the twentieth century. Initially, there was strong backlash by a fearful minority. Yet, the plan continued to install closed-circuit television (CCTV) cameras around the city. By 2011, more than 1.85 million cameras had been erected throughout the United Kingdom. 11,000 specifically cover the London Tube, with many others placed in public areas and near apartment complexes, overseen primarily by private companies.[216] Abuse by those with access or in political power certainly remains possible. Nevertheless, intellectual and political debates in the UK continue to be at least as fervent as before. The major political parties routinely reshuffle their proportions in Parliament, while arts and recreational sports remain popular sources of entertainment, as before. Additionally, in the press and by the public, the government is still

---

[215] See, for instance, Cartwright, "Cohortes Urbanae."

[216] Lewis, "You're being watched: there's one CCTV camera for every 32 people in UK."

criticized for serious errors it makes.[217] On the whole, an increased state of surveillance in London has not at all stunted robust conversation. In fact, the surveillance society appropriately enacted may even invite greater dialogue and interaction by building a safer space.

Undoubtedly, complete opacity of government surveillance opens the door wide for abuse. However, there is much middle ground whereby internal government organizations can conduct audits. The public may also need to have a right to question particular cases so as to ensure that vast power is not abused. Certainly, the benefits of surveillance are immense, and these are only some of the benefits drones offer.

SPOTLIGHT: HOW CAN DRONES ACTUALLY SAVE LIVES?

The myriad uses of drones include further life-saving potential. In June of 2017, Swedish researchers revealed their successful efforts to deliver automated external defibrillators (AEDs) to various locations, and to do so at speeds consistently faster than regular ambulances. In the United States alone, for the 350,000 non-hospital cardiac arrests that occur annually, 90% prove fatal (Columbus, "Could Drones Help Save People in Cardiac Arrest?"). Drone-delivered AEDs could substantially help save lives. If coupled with E911 smart-device services as addressed in Chapter 2, the results could be even more powerful. Of course, certain privacy concerns might still remain.

Economic, Social, and Other Benefits of Drones

Beyond bolstering security, UAVs' unique abilities promise to also boost knowledge and opportunities for all. Financially, drones stand to benefit the United States (in addition to the world economy)

[217] For instance, see Rawlinson and Greenfield, "UK government accused of 'pathetic' lack of help for Anguillans hit by Irma," criticizing delayed relief efforts for September 2017's Hurricane Irma.

enormously. As the Obama White House projected in 2016, the drone industry can inject up to $82 billion into the economy, while creating around one hundred thousand jobs within a dozen years.[218] A separate 2016 estimate conducted by the FAA predicts within 2.99 and 4 million total drones, commercial and hobbyist, to inhabit US skies by 2021.[219] Through the fall of 2017, these numbers seem to be on track with continued investment as several large corporations have each purchased approximately one drone startup per month.[220] Hence, the market is highly lucrative, and to continue being a player on the world's technological scene, the United States should continue to invest as well.

Underlying this financial value is the fundamental merit of drone technology. The assets that make them valuable for surveillance, also make them superlative for other tasks. Drones have already proven their worth in search-and-rescue missions, such as rapidly locating an elderly woman who had lost herself in a North Carolina crop field. On a day when temperatures reached 42°F extended exposure could have been fatal if not for deployment of the sheriff's office drone.[221] Drones are also being designed to assist firefighters in multiple aspects of their work. Following the 2017 London Grenfell Tower fire, for example, drones helped to survey for structural damage. The Fire Department of New York has used drones to search for trapped building inhabitants and

---

[218] Hood, "FAA Gives Commercial Drones Takeoff Clearance."

[219] FAA, "FAA Aerospace Forecast: Fiscal Years 2017-2037."

[220] Calhoun, "3 Reasons Why the Drone Industry Is Hiring Thousands of New Workers."

[221] Fox News, "Drone used to find missing woman in North Carolina cornfield, video shows."

identify points of extraction. Moreover, Dubai's DigiRobotics has developed a UAV that emits a fire-fighting foam.[222]

Commercially, drones have many uses as well—they can be used by both professional and amateur filmmakers to capture cost-effective aerial or other shots that might otherwise be prohibitively expensive.[223] Additionally, drones have scientific promise. They can soar to easily collect atmospheric measurements, and small, disguised drones have also been deployed in Antarctica to weigh seals and hence check on the ecosystem's health without disturbing the creatures.[224] As still another exciting development, California drone company Zipline launched a partnership with the Rwandan government in October 2016 to deliver blood to residents. In western Rwanda, many roads are sparse and unfinished with residents hence isolated. With their aviation capabilities, drones promise goods to individuals in such secluded circumstances with high precision, and under almost any weather conditions, day or night.[225] Undoubtedly, drones have life-saving potential.

There are truly boundless possible applications of value for drones. Public safety, scientific exploration, medicine, and other professional or hobbyist endeavors are only some of the areas in which drones will land. Encouraging innovation means that new problems can be tackled and solved. Investment also means that drones can continue to evolve. Consider, for instance, the Aerovironment Hummingbird drone. Through DARPA investment over just two years (2009-2011),

---

[222] For more on these fire-fighting drones, in order, see Kutner, "London Firefighters Used Drone to Battle Grenfell Tower Blaze;" Jansen, "NYC firefighters use drones to help battle blaze for first time;" and Nasir, "Firefighting drone to put out fires in Dubai in the future."

[223] See, for instance, drones compatible with the petite but powerful GoPro camera, as in Young, "Best Drone for GoPro."

[224] See Murison, "Drones are Helping Scientists Weigh Seals in Antarctica."

[225] Zipline. "How Zipline Works."

researchers improved the UAV's flight time from merely twenty seconds to eight minutes.[226] Further modifications to improve resilience and navigation indoors continue to improve its abilities.

Investment in drones will certainly yield dividends. This investment will mean recognizing some of the safety and perhaps privacy risks involved with the unmanned vehicles. Yet, it should also be clear that drones' risks are typically overstated. By encouraging drone growth, we will bolster public safety, boost the economy, and boost opportunities. New additions to knowledge and lives saved will in turn beget further opportunities. There is simply too much to gain for us to let the opportunity fly by whilst binding innovation due to unsubstantiated fears. The rotors of positive change, drones will lead us to a better future.

A Case For Increased Drone Regulation

The promise of drones may gleam in the distance, but the technology's darker potential hangs as a storm cloud above. Drones pose a significant threat on numerous fronts. These include the danger of misdiagnosis, the potential for citizen abuse of the technology at the expense of others, and most significantly, the danger of abuse by government operatives. Ultimately, these numerous vectors for misuse substantially increase the risk of the Chilling Effect, whereby activity, conversation, and thought are stifled, shattering the very foundations of our democracy.

Abuse by Citizens

Similar to how algorithms can be appropriated by selfish "Google bombers," drones can also be used for ill by civilians against their peers. With their tiny frames and ability to camouflage, civilian models of drones can be used by the paparazzi, voyeurs, and stalkers to

---

[226] *Fox News Tech*, "Robot Hummingbird Spy Drone Flies for Eight Minutes."

go where uninvited. Scheming criminals as well, looking to scope out the security system of a target's home, have already been caught in England performing reconnaissance via drone; their drone was purchased for merely thirty pounds.[227] Undoubtedly, others have escaped detection. While in some cases the actions of these rogue drone users are clearly illegal, in others, such as when the paparazzi shoot footage from outside a celebrity's private airspace, the law remains unclear. This calls perhaps for high-profile prosecutions of violations to set the tone, but more importantly, better regulations and rules of use than are currently promulgated. If drones are allowed to simply roam free, they can continue to intrude upon the most basic form of personal privacy.

Accidental Government Misuse

Another vector for drone misuse arises without design. Law enforcement, even with the best intention for securing public safety, may run aground. As with overreliance on flawed data and algorithms, discussed in Chapter 4, law enforcement's drone use can also result in unjust action. In turn, false positives have the potential to senselessly hijack citizens' lives.

Currently, drones are deployed in high-stakes military operations abroad.[228] The facial-recognition and heat-signature technologies can identify targets with relatively high accuracy, error rates in the single digits.[229] Computer vision remains one of the hottest fields in computer science, and widespread efforts seek to improve computers' abilities to recognize faces and individuals. Nevertheless, one of the key differences between facial recognition abroad and at home is the scale. When used abroad, the use is usually highly focused with a particular match (e.g., a

[227] Barrett, "Burglars Use Drone Helicopters to Target Homes."

[228] Asaro, "Dr. Peter Asaro on Drone Technology in Eye in the Sky."

[229] Brown, "NIST: Performance of Facial Recognition Software Continues to Improve."

known terror suspect) already in mind. This match would have a clear photo uploaded by the authorities with all the relevant features visible (e.g., with location of ears, cheekbones, and hairstyle all readable by the sorting program).

When used in public spaces domestically, however, as in the 2016 NFL Super Bowl,[230] there are more individuals under consideration and inevitably more unknowns. We could reasonably expect this to result in higher error rates, and research has borne this out. One 2016 study, for instance, found that a facial recognition algorithm ostensibly accurate to 99% with training sets, in fact dropped below 75% accuracy when faced with scales representative of a major city.[231]

If you consider the more than 70,000 attendees to Santa Clara's Levi's Stadium, an error rate of merely 1% would still mean more than 700 misdiagnoses. In the case of false negatives—undetected threats—the efficacy of the program would rightfully be under fire, given the financial and ongoing social costs. In the case of false positives, law enforcement would not only waste resources, but hassle innocent individuals and potentially even proceed with wrongful convictions.

Other unintended problems emerge from the conjunction of surveillance and artificial intelligence. Yet, if surveillance is to be performed in real-time and on a scale capable of preventing atrocities such as the 2013 Boston Marathon bombings, then automation is inevitable. Human operators, however sharp their vision, commitment, and reflexes, will struggle to stay as alert and perceptive as machines. Moreover, the scale of information to be collected—whether from the

---

[230] Glaser, "If You Go Near the Super Bowl, You Will Be Surveilled Hard." Drones have become a staple in subsequent football events, though deployed in other capacities. In 2017, Lady Gaga featured drones in a dazzling display: Burns, "Intel pwered the drones." In 2018, Minnesota law enforcement deployed drones to update their maps in preparation for the event: Chavey, "Be on your best behavior."

[231] Langston, "How well do facial recognition algorithms cope with a million strangers?"

thirty thousand drones projected to fly the skies or from just a few drones observing a city center—transcends human abilities. Algorithms employed by artificial intelligence offer an impressive solution to this, but they also introduce opacity that human controllers may not be able to decipher. In fact, it is beginning to be widely accepted that the most accurate AI systems fundamentally frustrate human understanding.[232] Such opacity, as considered in Chapter 4, raises still other problems including the propagation of ill-founded biases that would contribute to wasted resources and worse.

Another serious issue with increased surveillance involves the life cycle of collected data. While autonomous drones in the near future may simply operate in real time, the information would be more valuable if stored than if immediately discarded. Such storage would be valuable as training data for future drone systems. It would also be valuable in itself as a historical record for future investigations or audits. However, stored media would inevitably create a database ripe for attention that would be extremely informative in the wrong hands,[233] potentially even more so than the location data discussed in Chapter 2.

Intentional Government Overreach

Perhaps most significantly, increased security efforts mean placing some level of blind faith in government operatives. Complete transparency is, of course, impossible, since then targets would have full knowledge of how to evade the authorities. Yet, even if the vast majority of officials are altogether trustworthy, just a single determined individual can undermine a whole system. President Nixon and his autocratic impulses are of course infamous, as are Senator McCarthy's mad cries of "Communist!" that filled the Capitol and affected at least thousands of lives. Other scandals of nepotism and corruption at the highest levels of

232 For instance, see Brynjolfsson and McAfee, "The Business of Artificial Intelligence."

233 For more on the dangers of databases, see Chapter 8.

government—the carpetbaggers during Reconstruction, the Teapot Domers under Warren Harding, CIA director Oliver North commissioning arms trades to Iranians—have occurred and their ilk are likely to continue. Hence, channeling power to a central spot only increases the temptations and risks to the entire nation.

Yet, this sort of channeling is precisely what has been occurring through the twentieth and twenty-first centuries. Executive power has steadily increased since President Franklin Roosevelt, authority divested from the collective of Congress and bequeathed to a single individual. In wartime, over short periods, increased power to the commander-in-chief may be sensible to enable coordination and rapid action. However, a recent resolution—The Authorization of Use of Military Force (AUMF)—has broadened the timeline for concentrated power, and remains in effect. Undermining democracy from within, it also has dangerous potential for domestic drone deployment by law enforcement.

The Danger of the AUMF

Presently, military drone strikes are considered "targeted killings," and their authorization is tied to the 2001 Authorization of Use of Military Force from the US Patriot Act. The AUMF authorizes the president to "use all necessary and appropriate force against those nations, organizations, or persons he determines planned, authorized, committed, or aided the terrorist attacks that occurred on September 11, 2001, or harbored such organizations or persons, in order to prevent any future acts of international terrorism against the United States by such nations, organizations, or persons."[234]

The AUMF was naturally used to legalize attacks on the al-Qaeda network responsible for the 2001 attacks. Even after the killing of al-Qaeda leader Osama bin Laden, however, the AUMF has continued to justify attacks on other terrorist cells including ISIS, a distinct

[234] For a link to the full AUMF text, see this chapter's "Further Exploration" section.

organization from al-Qaeda.[235] Although there have not yet been successful challenges to this justification in court, one is pending,[236] and moreover, many civil rights groups have criticized what seems to be unlimited authorization to the commander in chief. President Obama himself professed a desire to change America's "perpetual wartime footing,"[237] despite the former resolution still carrying legal heft.[238] Indeed, there has been bipartisan support, including from Senators Tim Kaine (D) and Lindsey Graham (R), for a new AUMF denoting ISIS and other networks as explicit targets of force. Thus far, however, efforts have failed to produce a new or clarified resolution in Congress.[239]

This "perpetual wartime footing" is wasteful. In granting nearly unlimited power to engage in conflicts and send in troops, the perpetual war mentality can draw us "into more wars we don't need to fight."[240] Domestically as well, the implications are wide. The AUMF declares a time of war, and in wartime, civil liberties of citizens traditionally have been curtailed. Whether suspension of habeas corpus (the right to a timely and fair trial) by President Lincoln for POWs during the Civil War[241] or the internment of Japanese Americans during World War II, such devastating moves are controversial but nonetheless legally permissible when the country is engaged in conflict.

In fact, one of the initial justifications for domestic surveillance

---

[235] Sly, "Al-Qaeda Disavows Any Ties with Radical Islamist ISIS Group."

[236] ACLU, "*Doe v. Mattis* – Challenge to Detention of American by U.S. Military Abroad."

[237] White House, "Remarks by the President at the National Defense University."

[238] Acosta and Diamond, "Obama ISIS Fight Request Sent to Congress."

[239] McCauley, "Thanks to US Congress, Endless War Will...Not End."

[240] White House, "Remarks by the President at the National Defense University."

[241] Ewers, "Revoking Civil Liberties: Lincoln's Constitutional Dilemma."

programs offered the AUMF as a defense.[242] Some pieces of legislation, including the pivotal FISA of 1978 and the revamped 2015 USA Freedom Act, implement limited restrictions on US citizen surveillance.[243] However, the executive's wartime powers remain difficult to legally challenge, due to the president's "plenary" power in foreign affairs, as upheld by the Supreme Court.[244] Moreover, the legal designation of potential terrorists as "agents of foreign powers"—including those including those located within the United States—has been another tactic to accommodate FISA, and provide the Executive with wide leeway.

The capacity of ill-intentioned agents to strike violent, destructive blows is enormous, particularly with today's fast-improving technology. Undoubtedly, appropriate measures need to be taken to protect the security of national intelligence and military operations. Yet, citizen privacy and other civil liberties must also be defended. These are, after all, the core of our democracy. It may be the case that we have not yet descended into an Orwellian state. However, that is not an argument for continuing down the comfortable path of least resistance. The trouble with looking to history is that the pervasiveness of modern technology is unprecedented. Just as proponents of drone surveillance argue that terrible public attacks should be prevented rather than retroactively dealt with, so too should government abuse be addressed proactively. Otherwise, a single individual could overthrow the system to a point of no return.

---

[242] ACLU, "ACLU v. NSA."

[243] For instance, see Kelly, E., "Senate Approves USA Freedom Act."

[244] This notion of the president clothed in immense plenary power was explicitly supported in the 1936 Supreme Court case *US v. Curtiss-Wright*.

Cooling the Chilling Effect

Each of these misuses of drone technology are dangerous in themselves. All moreover contribute to a further and fundamental threat to society: the Chilling Effect. This "effect" has emerged as a legal concept in the late twentieth century, whereby free expression is stunted because of fear of surveillance. Since speech is intimately related to thought, the suppression of speech will lead to suppression of thought. Since thought inevitably affects actions, a Chilling Effect will impact individuals' autonomy.

SPOTLIGHT: HOW CAN CORPORATIONS BALANCE CONSUMERS' AND GOVERNMENTS' INTERESTS?

Should corporations comply with government surveillance demands? For most companies, they are caught between Scylla and Charybdis. On the one hand, obedience means government favor, while disobedience means possible prosecution, with hefty fines and possible criminal penalties. On the other hand, obedience may also mean widespread consumer disaffection in the future—and it may also mean drawing the ire of foreign governments. Some governments have even refused to engage with particular tech giants unless certain guarantees about their citizens' data are made and upheld.

In the wake of the Snowden revelations, companies such as Yahoo, Microsoft, Google, and telecom companies (despite earlier compliance) have resisted government requests (see Ackerman, "Microsoft, Facebook, Google and Yahoo Release US Surveillance Requests" and MacAskill, "Yahoo Files Lawsuit against NSA Over User Data Requests"). Fear of the public and for their business may well be driving the newer trend of transparency, thanks to the release of transparency reports and raw numbers of government requests for data through the previously secret National Security Letters that require no judicial review and the use of which was greatly expanded through the USA Patriot Act. In any case, this transparency trend is only one means that corporations have of rebuilding trust.

Numerous studies suggest that whether observed by a human or machine, most humans are uncomfortable with surveillance.[245] When watched, we are less likely to engage in behavior deviant from the norm. For violent, destructive behaviors, such suppression of impulses is perhaps commendable. However, for other sorts of atypical behaviors, conformity is dangerous. Unusual ideas are what make us unique, and drive beliefs of self-worth. Politically, unusual ideas are also the pride of democracy where the best ideas can be weighed against one another, and the voting public selects the most compelling. Progressively, unusual ideas are also what typically lead to innovation. Economically, innovation tends to feed more efficient ways of doing things, while driving scientific progress as well.

Importantly, simply the idea of surveillance is sufficient to disquiet individuals. Even if drones are not furnished with functional cameras, citizens who believe they are being watched are liable to restrict their actions. This will particularly be the case with the deployment of disguised drones. Anything in the natural world could then be viewed with reasonable suspicion—a squirrel or hummingbird might actually incorporate sensors, whether for use by the government or a personal stalker. It may seem hyperbolic, but the snowballing effects of drone surveillance could well launch an avalanche. The sanctity of the natural world—the sort appreciated by Thoreau, for instance—would be undermined with a folding of the artificial into the natural, and all for a nebulous notion of security.

It is not hyperbole to note that a culture of conformity fundamentally undermines democracy. Static "groupthink," as various scholars have argued, was likely a major contributor to past horrors including the Holocaust, the Salem Witch Trials, institutionalized

[245] See, e.g., Bruce Schneier, Data and Goliath, Chapter 7, and Schwartz and Solove, "The PII Problem: Privacy and a New Concept of Personally Identifiable Information," p. 1853, Which discuss numerous studies that have supported this.

slavery, and the Red Scare.[246] While London's implementation of CCTV cameras may have had no serious negative repercussions as of the present, the infrastructure remains in place for something nefarious. The vast majority of members in law enforcement may be trustworthy, like Christopher Nolan's Batman at the close of *The Dark Knight*. Yet, it only takes a few to overthrow the system for their own iniquitous ends. Other instances of surveillance, such as the East Berlin Stasi government and the People's Republic of China under Chairman Mao, indicate some of the terrifying consequences that were realized. They provide compelling evidence that blanket surveillance can indeed suppress speech, ideas, and action in horrible ways. Through opaque law enforcement activities—something that the AUMF may substantiate—the populace lived under these regimes not in safety, but in fear.[247]

These dangers are not as distant as they may appear. During the 2015 Freddie Gray march, for instance, a private company working with Baltimore Police used drones to capture a disturbing amount of detail about protestors.[248] Released footage showed how the drones zoomed in high definition to analyze particular individuals and vehicles over the course of the demonstration. Although no known prosecuting action was taken using the footage, doing so seems a logical next step. One recent case, *Carpenter v. United States* discussed in Chapter 2, is focused on the use of a suspect's location data. Based on his smartphone's

[246] For instance, in *Eichmann in Jerusalem: A Report on the Banality of Evil,* Hannah Arendt famously laid bare the hive mind that helped drive the Holocaust, with affirmation through echo chambers. This was not unique to Nazi Germany, and the same phenomenon could allow individuals in power to again implement horrific, dehumanizing programs.

[247] For an account of the Stasi surveillance regime, see Peter Wensierski, "East German Snitching Went Far Beyond the Stasi," and for an account of spying in China under Chairman Mao, see Branigan, "China's Cultural Revolution: son's guilt over the mother he sent to her death."

[248] Reel, "Secret Cameras Record Baltimore's Every Move from Above."

location coordinates, law enforcement associated Carpenter with a gang of electronic store thieves, and convicted him. If this sort of circumstantial linking of individuals continues, it increases the possibility of wrongful convictions moving forward: in a future protest turned violent, for example, a passerby in the march's vicinity might simply be heading to work, though her face captured by a surveilling drone; in turn, she could be convicted as a disturber of peace. Surely, that possibility, however remote, would weigh upon the minds of numerous citizens?

Consider for yourself: how might you feel if your face were under surveillance while shopping for groceries, or even when reading this book? Perhaps on the basis of your facial expressions at each page, you would be charged differently. Yawns of indifference might translate into refunds. Smirks or furrowed brows in concentration might become extra charges. Would you not change your behaviors in response? Would you not advise your family and friends to also do the same? On the whole, the hope of creating a safer society through comprehensive surveillance is deeply misguided.

Just two hundred years ago, "We the People" built this government. Today, an even more inclusive "We the People" must uphold it. America was founded to combat tyranny and enhance the citizenry's autonomy. Focused executive power and a perpetual declaration of war are simply unacceptable. An indecisive Congress must be held accountable; otherwise, America heads down a slippery slope that leads back the way it came.

Projecting drone technology's possibilities, several prescient public interest groups (e.g., EPIC and the ACLU) have long warned of privacy risks and other dangers. These voices must not drone on unheeded. The dangers of abuse are severe in themselves, as well as for the Chilling Effect the technology is liable to produce. While some level of surveillance may be necessary for security purposes, we should not simply accept a surveillance state. Strong oversight mechanisms absolutely must be in place.

SPOTLIGHT: HOW DEFINITE IS DRONE LAW?

In mid-2016, the FAA published a long list of guidelines for drone use, with the overt intention of making drone use more accessible. Various developments, however, have shown that the FAA rules are incomplete, if not insufficient. On May 19, 2017, a federal appeals court ruled that the FAA's requirement that drone owners register their vehicle is unconstitutional (see Vanian, "You No Longer Have to Register Your Drone"); as a result, there is at least a short window where individuals can purchase and use drones with minimal oversight.

Moreover, invasion-of-privacy cases have emerged with uncertainty over the proper legal designation of airspace. What is private property? What is public property? At what heights? On August 1, 2017, a local news station flew a drone over the new home of Kentucky Governor Matt Bevin, possibly filming his children and other residents without consent. While the Supreme Court case *United States v. Causby* established that airspace at least as high as 83 feet was private space, the applicability of the 1946 ruling today is unclear (Farivar, "Governor Angrily Tweets about Local TV Station Flying Drone over His House").

"The Drone Slayer," featured in another Kentucky case, raises further questions. Resolved earlier this year, this case featured a man named William Merideth, who had shot down and hence destroyed a drone he believed to be infringing upon his private residence. The drone owner had sued for damages, but the judge in the appeals court dismissed the case as a state affair, upholding the lower court ruling in Merideth's favor (Farivar, "Judge Rules"). The question of whether homeowners can shoot down drones without financial damages remains unresolved on a federal level.

In Conclusion

Domestic drone use offers great benefits as well as various risks to citizen privacy. On one hand, it offers a powerful way to prevent devastating attacks by organized terrorists and lone gunmen. It promises, as well, significant economic boosts while furthering innovation and knowledge. On the other hand, it introduces various risks to privacy.

Beneficial and harmful uses are the traits of any technology; it is in the implementation and the general standards of practice that the trends of use kick in. These are, in turn, influenced heavily by regulations, and it is here that some advocacy groups are concerned. Some privacy ideas are rooted in the practice of law. Likewise, some explicit cases such as *United States v. Causby* have established baseline regulations for hobbyist aviation. Yet, few regulations about drone usage have focused on privacy. Perhaps some early cases will establish the lay of the land, scoping out what is and what is not permitted; nevertheless, the bounteous possibilities of drones also open up new avenues and will continue to warrant new considerations.

In our world, with press freedoms often limited,[249] transparency does not see the light of day. Yet, transparency is key for any democratic government. Precisely where the demarcations will be remains to be determined. Some secrets will likely need to be exposed, but individual privacy rights will also have to be balanced with security. As states and the national government move forward with legislation impacting the drone market and use, we will have to raise our voices above the noise to shape the future as we desire. Whether that future consists of more regulations or affirms the status quo, our awareness as citizens is key.

[249] Freedom House, "Silencing the Messenger: Communication Apps Under Pressure."

Further Exploration

The 2016 Wiretap Report shows the rate of wiretaps approved by US courts: **http://www.uscourts.gov/news/2017/06/28/2016-wiretap-report-federal-orders-rise-state-totals-fall.**

Many **prominent pros and cons for military drone strikes** have been compiled at **https://drones.procon.org**. The arguments, with citations for further consideration, are curated by the "nonprofit nonpartisan" **ProCon.org**, which aspires to promote critical thinking, education, and informed citizenship "by presenting controversial issues in a straightforward, nonpartisan, and primarily pro-con format." Other topics covered include the value of social media and the importance of net neutrality.

The **ESRI Story Maps** team in collaboration with **PeaceTech Lab** has created a crowdsourced global map of terrorist attacks at **http://storymaps.esri.com/stories/terrorist-attacks**. The results include the number of fatalities and location of the attack and go back to 2016. To toggle an available year X, enter "?year=X" after the URL. Another valuable resource for viewing data on terrorism's global impact is the compilation "Terrorism" by Max Roser, Mohamed Nagdy, and Hannah Ritchie at **Our World in Data**: **https://ourworldindata.org/terrorism**.

A compilation of federal and state **laws governing commercial UAVs** in the United States can be found at **www.ncsl.org/research/transportation/current-unmanned-aircraft-state-law-landscape.aspx.**

For more on **government surveillance through means other than drones**, see Bruce Schneier's ***Data and Goliath*** and Frederick Lane's ***American Privacy***.

The **AUMF**, which justifies much of American military policy in the twentieth century, can be found at **https://www.congress.gov/bill/107th-congress/senate-joint-resolution/23**.

Gavin Hood's film *Eye in the Sky* dramatizes the political, legal, and moral issues embroiled in military drone use. It features an all-star cast including Helen Mirren, Alan Rickman, Aaron Paul, and Barkhad Abdi, and a taut script by Guy Hibbert. Numerous military and technical specialists have highlighted the film's accuracy. See, for instance, reviews in *Wired* (McFarland, "*Eye in the Sky* Is the Quintessential Modern War Film") and *The Guardian* (Gregory, "How a Hollywood Film Reveals the Reality of Drone Warfare") listed in References.

The prospects of bioterrorism may be frightening. Nevertheless, the United States' Center for Disease Control (**CDC**) has prepared accessible educational materials to arm citizens. **Fact sheets**, including information on possible treatments, are publicly available at **https://emergency.cdc.gov/bioterrorism/index.asp**

EPIC's "Spotlight on Surveillance" provides the latest on the organization's efforts to bolster consumer control over surveillance: **https://epic.org/privacy/surveillance/spotlight**.

An expanded discussion of the benefits and hazards for domestic drone use can be found in the **4PrivacyMatters** publication ***Wheels in the Sky: Drones and Citizen Privacy***, available at **https://www.amazon.com/Wheels-Sky-Drones-Citizen-Privacy-ebook/dp/B07CYZVYFF**.

Chapter 7: The Privacy of "Things"

Key Questions:

- What is the Internet of Things (IoT)?
- How large is the IoT today, and how large will it become?
- What are the dangers of a digitally interconnected world? What are the benefits to ordinary citizens as well as to corporations and governments?
- What can be understood from Big Data analysis? What are its limits?
- How can benefits and risks be communicated and enforced to prevent dangerous policies from coming to be?

Consider: Roommates José and Miguel live in a neighborhood known for its high-crime but low rent. To reduce their risk, the pair has just installed a home-security system that provides assorted home-protection services, relatively cheap. After some debate, they also placed a large sign outside their door that warns would-be intruders of the security inside. Among other services, the system automatically alarms doors when all of a home's marked residents leave, using retina scans to deactivate the alarm upon reentry. Moreover, it has an interface to identify and immediately highlight suspicious activity: for instance, if any doors are left ajar for greater than two minutes, they will be illuminated and trigger a notice. Through the cameras spread throughout the house, a homeowner can view activity in any room and also conduct video calls with residents indoors.

Besides homeowner vision, the cameras are further monitored in the cloud by an artificial intelligence program. This program checks for sudden movements and uses facial-recognition technology to identify residents inside the home at any moment, in order to note whether stray animals or would-be robbers have entered the residence.

In the past few weeks, a few false alarms that mistook rodents for robbers forced Miguel to rush back to the apartment, and he became increasingly annoyed with the system. He was also flustered to occasionally note the red light that indicated an idle watching status. He called the security company, which apologized for the mistakes and explained that the system often had a learning curve as it acclimated to a new environment. Moreover, the company representative stated that the camera had to idle in order to stay vigilant—home footage, however, would never be accessed by human employees unless at the request of the homeowners. This was strict company policy, though Miguel was not wholly convinced.

Just a few days later, the alarm accurately identified a stray cat that had attempted to slink in through the window. A few days following, a series of neighboring apartments were violently vandalized, and José suspects the reputation and efficacy of the security system helped to defend his own apartment.

It has now been a month with no false alarms in the interim and increased crime throughout the neighborhood. However, the company has also released a statement stating that they will now have to increase the cost of services for all customers as part of the upkeep costs of increased global data attacks. As the roommates decide about the future security of the apartment, José wants to further upgrade the system, while Miguel remains ever-suspicious. Is Miguel's concern about being spied upon a legitimate one? Does the behavioral data collected from idle watching serve a justifiable purpose? Is the company abusive for increasing prices for customers already hooked in?

—Inspired by current home-security systems

Lest you need any more proof about the modern world's complexity, consider that *Thing* is now a technical term. And, as with many pervasive terms, the slimy Thing evades easy definition. In broad strokes, Things are typically thought of today in relation with a network termed the "Internet of Things" (IoT). Named somewhat deceptively, this network is not a single network but includes localized and isolated networks in addition to global ones. Collectively, however, the IoT does consist of interconnected electronic devices that collect and exchange data with one another. Member devices, like members of the human species, include such diverse Things as wearables (e.g., Fitbits or Apple Watches), smartphones, tablets, temperature sensors, and home-security systems, like the one purchased by José and Miguel.

In general, each IoT device consists of three main elements: the actual device or sensor (such as a home-security camera or series of cameras), supporting applications for accessing or controlling data (such as a mobile or computer app that displays the device's or sensor's readings and evaluations), and back-end or cloud services (such as data stored, perhaps analyzed in the cloud) that make the data easier to digest and accessible for review. Running on data and basic computation, IoT devices have multiplied massively in recent years, in conjunction with microprocessors' reduction in size and cost. We already considered the growth of one type of IoT device—the smartphone—in Chapter 2. Other smart devices such as home energy and temperature sensors are also propagating. In fact, numerous independent estimates project that within ten years the overall market for Things will blossom to several trillion dollars and twenty billion or more total IoT devices.[250]

Even as consumer adoption of Things grows, industrial Thing use is expanding even faster. Already, many of these industrial Things control basic utilities. For instance, rudimentary Things are responsible for managing the air quality in offices, controlling turbine speeds in power plants, adjusting light intensity in theaters, and modifying paths

[250] For a compliation of estimates, see Postscapes, "IoT Market Forecasts."

of robot activity in factories. Moreover, analysts estimate that by 2021, market dynamics will shift such that industrial Things dominate, accounting for 58 percent of all connected devices.[251] By then, the societal impact of Things—their benefits and their risks—will be wide.

Some speed bumps have slowed growth from near-term projections. Accenture recently found, for instance, that across twenty-eight countries including the world's ten biggest economies, the spread of consumer Things is slowing, contrary to expectations. Key inhibitors inferred were excessive prices and consumer concern with security.[252] Neither of these concerns is indomitable, but their emergence also indicates that the IoT remains in development. While the long-term prospects of the IoT may be as strong as before, Things' success may well depend on explicitly addressing these concerns.

The home-security system described at this chapter's start combines a variety of issues related to privacy and modern technologies. The system performs a critical function for its owners, but does so via surveillance in an intimate realm: the home. Moreover, the system performs this function through analysis tools in the cloud, tools that also leverage the philosophies of Big Data. Not all Things are equally complex or well-connected, but by combining several smart devices, José and Miguel's home-security system is among the best—and perhaps most dangerous—that the IoT currently has to offer. Different responses to the common consumer concerns above will be addressed in the perspectives below.

A Case For Rapid Expansion of Things

With its promise of tremendous benefits, the Internet of Things must be allowed to flourish through rapid expansion. The risks posed

[251] Sealy and Menting, *IoT Security and Managed Services.*

[252] Accenture, "Igniting Growth in Consumer Technology."

are no more dangerous than those for any other technology and can be adequately addressed to minimize their impact. Some of the IoT's chief benefits include the massive efficiency it promises and has already begun to deliver, the interconnected ecosystem it leads towards, and the phenomenal value provided by the Big Data which both feeds and is fed by it. At its core, uninhibited growth will also maximize innovation, activating still other benefits.

Massive Efficiency: Time = Money

IoT devices promise increased efficiency of our time and energy. While the devices may be experiencing some growing pains, their efficiency on the whole will translate out into many other benefits including increased security, improved knowledge, and increased potential for profit.

In a critical vein, a 2015 article published in *The Atlantic* questions the necessity of many Things prolifically marketed.[253] Citing the infomercial quality of certain services (e.g., smart gas-tank meters and smart bike locks that address already-resolved problems), the author criticizes a blind move toward nominal efficiency and a counterfeit ease of mind. While this critique may be deserved for certain products, consumption is driven by consumer desires and needs; what may seem unimportant to some may, in fact, be important to others. Nevertheless, given that at present, many of these excessively priced goods for unnecessary services are not selling well[254] across the globe, consumers, it would seem, concur.

Yet, the failure of some Things does not reflect on the technology's core potential or prospects. It is only natural that a large market which enables copious experimentation includes duds. The items truly of value for consumers have yet to be clearly identified and focused

[253] Bogost, "The Internet of Things You Don't Really Need."

[254] Accenture, "Igniting Growth in Consumer Technology."

upon. Instead, variety reigns today, especially for consumer Things.[255] More significantly, not all professed efficiency is equal. Efficiency of time may be overstated for some individuals but significant for others. A home gas tank gauge accessible by a smartphone app that takes several minutes to buffer, provides no real increase of efficiency: the owner could check the meter more quickly by simply walking up to the meter, and do so with significant enough precision.

Instead, the most successful products are bound to be those that innovate and produce something new. This can be something with greater functionality than before, greater precision, or greater ease of use. When the iPhone began the smartphone revolution, it did so not because it was the very first of its species—Nokia and Sony Ericsson among others had already beaten Apple to the design game.[256] Rather, it succeeded because of cultivated qualities. Through the meticulous oversight of Steve Jobs, it was ensured that the most frustrating problem of other companies' models was rectified—namely, an unwieldy and inefficient screen that obstructed user movement. Among other key changes, Apple removed the physical keyboard that occupied half the potential screen area of competitors' devices. Apple also designed the entire screen as touch-sensitive, allowing the sizable number of tactile learners to fully embrace their strengths. These innovations, not merely novel but fundamentally valuable, laid the groundwork for future smartphone models by all companies.

Similarly, the ongoing revolution of Things is an exciting moment of opportunity. There is great potential for what products can and will be produced. While gas tank meters with minimal value may continue to be designed in the short-term, there are many other products

---

[255] Postscapes, "IoT Devices & Products" identifies many of the most popular IoT products as of June 2018. Their breadth remains at least as vast as ever, with smart locks, key rings, electric plugs, sprinkler systems, vents, and much more.

[256] As comparison, one informal compendium of popular phones before the iPhone's launch is at H., "The phones before the iPhone: 10 iconic devices."

of undeniable value. For instance, wearable devices to track tremors in Parkinson's patients have yielded new insights into the disease while forestalling serious attacks through the monitoring of family members and medical professionals.[257] Also for health care, Fitbits allow users to closely monitor exercise, heart rate, and diet so as to improve or maintain their health. Smart fridges and other meters, commissioned by the UK for implementation by 2020, look to substantially improve energy efficiency by revealing energy leakage.[258] In manufacturing, the powerful GE Predix platform improves energy and time efficiency of production, while helping to forestall costly problems.[259] For home security, Canary offers a device like the one outlined in this chapter's opening, a product that monitors home security. If so enabled, its video cameras also allow video-conferencing with home inhabitants. Through its valuable connectivity, the system has been successful in its primary goal: it has helped to stop burglars and recover stolen items more quickly than would have been possible otherwise.[260]

For more leisurely consumer pursuits, the IoT provides other novel and valuable opportunities. Amazon Echo's Alexa permits easy discovery and playback of songs or radio via Amazon Cloud, Spotify, or other music services. Connected to the cloud, Echo and similar products, such as new competitor Google Home, allow users to increase efficiency in still other ways, for instance, allowing users to check their calendar or flight status verbally and enabling vocal web browsing. These connected devices offer increased time efficiency by circumventing manual queries and furthermore empower easier acquisition of knowledge.

[257] Giuffrida, "Wearables and Application Solutions for Parkinson's Disease."

[258] Smart Energy GB, "About the Rollout."

[259] Annunziata, "The Value of interconnectedness."

[260] Canary, "Burglars Caught by Canary."

Ultimately, the IoT should not be judged by its least successful products when the market is just beginning. Rather, the most successful products hint at where the market is primed to go and which products will last. Some of these innovative ones have already seized general interest, while others will soon do so. The will of the public is powerful and keen—it determines what is of true value, and we should trust its judgment.

Today, more than ever, time means money. Yet, along with the vast amount of information at our fingertips, there is also a vast amount of noise and distraction. While some Things may contribute to clutter, the best Things will help us to avoid distractions. Precisely how we use the added time, knowledge, and peace of mind that these products produce will be up to each of us. For its part, the home-security system purchased by Miguel and José is of a meaningful species that should provide its owners net savings and ease of mind.

The Big Data Boost

New Things are valuable and exciting for their own sake. Many of them—for instance, smart glasses by which our knowledge of our surroundings is enriched with historical and other relevant information—are even realizations of fascinating technologies envisioned in such classic science fiction paradigms as *Blade Runner* and *Minority Report.* As such, they push our world closer to an ideal of deep knowledge and productivity.

Things lead us to a better future through another means as well: their compilation and use of Big Data. Typically, the term describes data of large volume and large variety that arrives at high velocity, these relative terms intentionally left ambiguous to encourage still another common characteristic: flexibility. "Big Data" was something we considered in terms of informing algorithms in Chapter 4, and already, many IoT devices, such as smart thermometers and smartphones that rarely turn off, collect data in line with these parameters. For a consumer, this can be valuable, such as when a health monitoring device analyzes a user's sleeping patterns based on heart rate and recorded motion. Collected over a wider range, Big Data analytics offer still more insight into societal trends, potentially even shining a light on the essence of human nature.

As with the IoT, some have quickly dismissed the prospects of Big Data, citing numerous unsuccessful projects that lacked clear focus and failed to correct for grievous instrument errors.[261] However, when properly performed, Big Data analysis can provide groundbreaking results. *Moneyball*, based on events from the early 2000s, documented one of the first success cases for Big Data's use that arguably revolutionized sports,[262] ushering in a new team-centric era. The analytics approach continues to be used and has seen success in the dynastic Golden State Warriors,[263] the Premier League teams in the UK,[264] and baseball's Oakland Athletics with whom 2002 General Manager Billy Beane began the trend. These teams' front offices used data analytics to help identify valuable individual prospects in the draft as well as in free agency.[265]

Things, fed by Big Data and collecting it, build on the revelations. For instance, in sports, though fitness, equipment, and training techniques have developed to augment some of the strongest yet most agile human beings, serious injury risks remain. Through sensors and biometric data monitors, athletes such as Warriors point guard Steph Curry and Athletics first baseman Scott Hatteberg, both of whom were highly prone to injury early in their careers, have been able to compete splendidly. They can better assess the wear and tear upon

---

261 See, for instance, Ranger, "Big Data: Why the Boom Is Already Over" and Mepham, "The Deadzone: Why Big Data Is Dead."

262 *ZDNet*, "Sports Analytics."

263 Ross, "Welcome to Smarter Basketball."

264 Marr, "How Big Data and Analytics Are Changing Football."

265 Ross, "Welcome to Smarter Basketball."

their bodies, while optimizing their movements to lengthen their careers, improve their lethal skills, and better enjoy the game themselves.

As Lewis chronicles the movement in *Moneyball*, Beane, with analyst Paul DePodesta and his team, originally performed their novel data analysis out of necessity; they had limited finances and were vastly underfunded in comparison to their peers. Yet, the technique developed into much more than a cost-saving measure: it became a way to assess players' value more accurately. Whereas scouts may have ingrained prejudices about how a pitcher should throw and how or when three-point shots should be launched, the raw numbers offered objective insights. Instead of relying on old traditions merely because of convention or even because they offered success in prior eras, the Big Data approach has allowed for innovations suited to the modern game. Compared to the past, players in Major League Baseball have come to be valued in fundamentally different ways for their ability to "get on base," while NBA teams, following the Warriors, have widely adopted mass shooting of three-pointers.[266] Big Data does not tell a full story—undoubtedly, its insights must be combined with traditional wisdom as well—but it does allow for clearer windows into how things are. This clearer vision translates into more equitable treatment of individuals, who can now be considered on their own merits rather than blind conformity.

[266] In the NBA, the trend towards the three-point shot is clear with numerous related records broken in the last few years. For instance, the Cleveland Cavaliers set a new single-game record for made threes in 2017 (McMenamin, "Cavaliers make 25 3-pointers"), while the 2017-18 Houston Rockets, for the second consecutive season, set a new record for made threes in the regular season (NBA, "Once again, Houston Rockets set single-season record"). Some critics have questioned the viability of the three-point trend during the postseason when the games matter most, especially in the wake of two deciding 2018 Conference Final Championship games in which challengers to the throne died by historically futile efforts (e.g., Kram, "The Postseason of Our 3-Point Discontent"). Nevertheless, with the Warriors' ultimate success, a revolution does seem to be underway.

In many other fields as well, Big Data affords greater opportunities. Just recently, Big Data research has helped to identify disease biomarkers as well as substantially improve measures of financial behavior.[267] Also of great value in network analysis, Big Data has moreover helped to curtail the spread of disease and identify terrorist cells.[268]

SPOTLIGHT: YOU CAN PLAY WITH BIG DATA, TOO!

Big Data is not reserved for wealthy and established corporations—tools exist for all of us to use it. One platform for use is KNIME which is open source, free to download, and offers various tutorials as well: https://www.knime.org.

Depending on your objectives, data sets for analysis may be best collected fresh and under your own careful design. Many valuable sets, however, are available online. For health data, the NCBI database features scores of public gene information culled from research (https://www.ncbi.nlm.nih.gov/gene). Kaggle (https://www.kaggle.com/datasets) contains repositories of widely varied data sets posted by users. Still other data, such as tweets, can be automatically collected by setting up an appropriate program to capture data of interest. Big Data is typically characterized by great volume, velocity, and variety, so any such data set can be used to explore Big Data.

---

267 For more on biomarker use, see, for instance, Sekar et al, "Schizophrenia Risk from Complex Variation of Complement Component 4." Singh, Bozkaya, and Pentland, "Money Walks" discusses how the authors developed a better model of financial behavior through Big Data.

268 Christakis and Fowler, "Social Network Sensors for Early Detection of Contagious Outbreaks" addresses network analysis in predicting and preventing disease. Engelen, "Big Data Will Effectively Fight Terrorism in the World" addresses network analysis in determining terrorist cells.

Given these valuable uses, it will clearly not just be corporations in pursuit of better marketing that benefit from behavioral Big Data. Instead, communities and individuals stand to grow immensely as well. Hence, as we simultaneously balance consumer rights to information privacy, there are broad societal benefits bound to be unlocked via the Big Data compiled from Things.

Smart Cities and the Interconnected Dream

Valuable for its immediate impact on efficiency and innovation, the IoT is also integral to a magnificent longer-term vision: wholesale interconnectivity among both devices and people. This "connectivity" will manifest in several different ways.

For one, connectivity means even more efficiency. For instance, on local scales, networks of Things have helped to minimize information waste. As GE has noted, "every machine or controller might generate megabytes of data that often are not used to their fullest extent." However, through an interconnected network, all relevant data can be coordinated, integrated, and fused, which GE has done with its industrial network, Predix.[269]

Connectivity also means increased dialogue. Through smartphones, for instance, we have new ways to be connected and inform. We can connect with friends and strangers halfway across the globe almost instantaneously. We can text back and forth, comment rapidly on relevant political or recreational developments, and video chat so that, like time travelers, we fold the fabric of space-time to make distance disappear. On an even larger scale than what we presently have, this means that the whole world—even remote areas—will be merely a device away. Academic and business collaborations stand to benefit, as do dispute resolutions, and perhaps even political discourse. In turn, this exposure promises to increase inclusivity and reduce discrimination, due to greater exposure to humanity's dazzling diversity.

[269] For more on Predix, see Annunziata, "The Value of interconnectedness."

Furthermore, connectivity means increased security and access by relevant parties. One facet of this is security for at-risk individuals, such as family members with Parkinson's whose status can now be remotely observed. On the scale of a smart city, this also means that society can be monitored more holistically so that threats are public safety threats are addressed before they erupt. This may come in the form of automated drones, as we considered in Chapter 6. It may also come in other forms, such as surveillance of traffic flow via pressure sensors at crowded intersections that could help redirect vehicles and minimize logjams and possible road rage.[270]

As leading IT company NEC has noted,[271] increased connectivity may also mean that cyberattacks are easier to launch and perhaps even more impactful. After all, if one element of an interconnected system is hit, many others are likely to be affected. Moreover, with more connected devices come more potential areas of entry. Yet, these concerns about digital security, though significant, are hardly damning, for these threats are addressable. For one, some basic protections are simple, such as improved education about the vulnerabilities of Things and how to easily circumvent these risks (see next chapter's "Boosting Cybersecurity Via the Cloud"). Still other difficulties could be substantially rectified via adequate leadership.

Within the security industries, there is a sense of insufficient expertise at present.[272] In general, senior corporate executives and government officials appear to lack the cross-sector awareness needed to implement an adequate approach to security. If true, this offers another significant way by which to bolster the current ecosystem.

---

[270] For more detail on one particular method to monitor intersections—Solid State LiDAR (SSL)—see LeddarTech Inc., "Smart Cities—Using Solid State LiDAR Sensors for Traffic Management."

[271] *The Economist,* "Securing the Digital City: Cyber-Threats and Responses."

[272] Ibid.

Focused solutions may have short-term benefits for quickly resolving an issue, but unless broader solutions are developed to patch up all the holes in a network and educate all the responsible parties, vulnerabilities will persist. Intriguingly, then, interconnected communications and leadership can not only be the result of developing the IoT but in fact the security solution themselves. Improved channels of communication and data acquisition via the IoT can be translated into stronger security policies. Furthermore, the focus of cybersecurity can shift to improve defense. With so many devices uploaded, the concern may not be building robust defenses but rather proactively detecting threat sources and focusing attention on monitoring if not eliminating these.

While cyberattack news tends to dominate, security has always been a game of cat-and-mouse. Through social engineering and technological innovation, people have always been motivated to enter what seemed impenetrable. The legend of Robin Hood persists as an inspiring fairy tales for children. He may have stolen scores of apples, but he did so from rotten rulers and for a just cause. Butch Cassidy, Jesse James, and other antiheroes of the Wild West live on in tales as well. In cyberspace, hacking attempts also persist, such as Kevin Mitnick's hack of the Department of Justice server in the early 1990s.[273] More often than not these hackers and robbers have been caught and publicly disciplined; nonetheless, others, perhaps motivated by desperate circumstances or the thrill of illegal action, retread their footsteps.

Of course, the future will not necessarily echo the past; efforts to secure systems must continue. Researchers' ability to rather swiftly shut down the WannaCry ransomware in May 2017 attests to success. So too do does continued investment in Thing cybersecurity, as discussed further in Chapter 8. We can also grow faith from how banks and financial markets as a whole have been able to maintain security at such a high level despite the large number of players involved.

[273] For Mitnick's own account of his exploits see *The Art of Deception.* The work also includes case studies of other bold attempts at hacking important systems.

All in all, cyber-vulnerabilities are a significant concern but are not reason to discard the interconnected dream. The interconnected world is not a distant fantasy—already, cities including Singapore, Barcelona, London, Oslo, and Chicago have made strides to transform their infrastructure via the IoT.[274] Given the myriad benefits at hand, we should hasten this transformation, and the best way to do so will be through minimal regulation. Movement in the opposite direction would cripple innovation and the bounteous benefits bound to emerge from the market. Implementation of such regulations would require strong proof that the current ecosystem is in danger, but as will be discussed further in the second half of Chapter 8, cyber-vulnerability risks are vastly overstated. Moreover, the substantial benefits are still growing.

With a focus on cybersecurity of Things—and perhaps also a focus on devices that add legitimate value—the IoT can still strive for its elegant potential. Coupled with Big Data, the IoT will offer enormous value to consumers, businesses, and governments alike. Still in its infancy, the IoT should not have its future constrained by excessive regulations. We should trust traditional market forces to produce the products we desire and deserve.

A Case For Increased Regulation of the IoT

As with most things, acceptance of the IoT is not all-or-nothing. The options are not merely to continue production as is or cease it altogether. However, there are very serious issues with the current state of the Internet of Things. Until this is firmly acknowledged and incorporated into policy, growth must be halted.

Primarily, this stems from cybersecurity vulnerabilities that Things introduce and by which they have already been exploited to great

[274] For more on smart cities and current developments, see Buntz, "The World's 5 Smartest Cities," Bergen, "Chicago seeking 'smart-city' tech solutions to improve city life," and the "Further Exploration" section of this chapter.

effect. Perhaps some easy solutions are available, but the fact they have not yet been implemented indicates that they are not as easy as they sound. Even beyond these vulnerabilities, other elements of the IoT remain troubling. These include the tremendous potential for harm possessed by unsound Big Data use, as well as the darker implications of the interconnected dream, by which privacy could well grow endangered. These concerns appear to cut to the heart of why IoT growth has slowed of late, and indicate that through regulations, progress of Things should be curtailed.

The Great Vulnerability of Things

In 2013, former director of the NSA John McConnell claimed that ninety percent of the world's computers holding strategic, monetary, or intellectual value may have been infected by malware.[275] Of course, the actual percentage is impossible to determine with genuine precision, but even a small portion of devices, if infected, would have a domino effect. New products developed with these systems—including the software for most Things—would also be compromised. Moreover, any devices connected with previously infected devices could themselves become compromised.

This may seem like a speculative chain, but we do know of actual cases by which dormant malware unleashed catastrophic effects. For instance, the computer worm Stuxnet was introduced into an Iranian nuclear facility, presumably by the CIA in the early 2000s.[276] It dwelt inert for many years before activating and forcing the plant's shutdown. Though the full destructive potential was not released, the worm could easily have been used to turn off oil pipelines, modify pressure at the plant, and ultimately begin nuclear meltdown at the facility.

---

[275] NEC, "Ensuring Digital Safety with Robust Cyber Security."

[276] Nakashima and Warrick, "Stuxnet Was Work of US and Israeli Experts, Officials Say."

Perhaps it seems that this sort of power will always be in just hands. Setting aside the possibility of abuse by determined individuals as addressed in the previous chapter, this already is not the case. So-called zero-day vulnerabilities are loopholes or other dangers in popular software that are typically unknown to the public as well as the developers themselves. These vulnerabilities have a thriving market, sold for hundreds of thousands of dollars, either directly to interested buyers or on bidding sites. As insiders note, discovery of the vulnerabilities typically comes from state-sponsored teams. Not all these teams are allied with American interests, however, and once sold, these vulnerabilities are there for the exploiting. Governments engaged in cyberwarfare can exploit them, as can wealthy independent buyers with their own nefarious agendas.[277]

Possibly hidden vulnerabilities are one vector for exploitation when it comes to Things; the devices' sheer number is another. One type of attack critically relies on quantity: a distributed denial of service (DDoS). The largest attack yet occurred in October 2016, ten days prior to Halloween. Targeted was domain name system provider Dyn, and the successful attack shut down Internet service for much of the northeast United States during the day. A DDoS attack might typically be seen as more an irritant than deeply invasive, for it precludes users from accessing the bounty of the Internet rather than directly stealing any personal information.[278] Nevertheless, it can have a severe economic impact by shutting down business, and moreover restricting access to emergency services. In the forensics that Dyn ran after the event, the company estimated that "100,000 malicious endpoints" were responsible: that is, thousands of devices sent access requests to the Dyn servers in order to overload and shut it down. The culprit was identified

[277] For some more on "zero-day" exploits, see Patterson, "Zero-day exploits: A cheat sheet for professionals."

[278] Some more general information on DDoS attacks can be found at DigitalAttackMap, "What Is a DDoS Attack?"

as the Mirai botnet, which infected unsecured devices owned by otherwise-innocent consumers, and transformed them into malicious zombies that could be weaponized on command. Infected devices included digital cameras, DVR players, and smart thermometers.[279] Through the sheer number of devices poorly secured, the perpetrators were able to carry out their attack, and if the IoT continues to grow irresponsibly, there will be little means of stopping future attacks. Yet, though some solutions may appear to be basic, they have yet to be implemented.

After severe cyberattacks, security experts tend to release public "lessons learned." For instance, shortly after the Dyn DDoS Attack, experts recommended that users should immediately change the default passwords of IoT devices from factory presets. Towards mid-2017, another DDoS attack—the BrickerBot—launched. The key takeaway from this assault was to change devices from factory default login information.[280] Another prominent attack that struck in 2017 came as a result of hospitals failing to update their computers with a software patch that had been released a month prior.[281]

These post-mortems make the solutions appear rather simple, yet if they are so simple, why have they not been implemented? At least part of the answer appears to be that security has not been a priority. Perhaps this can be addressed through increased education efforts by Thing-producing corporations and improved device interfaces that make password changes easier to perform. Nevertheless, given the apparent

---

[279] Woolf, "DDoS attack that disrupted internet was largest of its kind in history, experts say" and Hautala, "Why it was so easy to hack the cameras that took down the web" provide some more information on the Dyn attack. More discussion of this incident and of other sorts of cyberattacks is featured in Chapter 8.

[280] See, for instance, Wallen, "Five nightmarish attacks that show the risks of IoT security."

[281] More on the WannaCry ransomware attack is in Chapter 8.

current lack of concern, merely continuing on with development will not suffice.

More broadly, possessing a unique IP address searchable through the vast Internet, many IoT devices remain accessible and hence vulnerable.[282] Moreover, besides DDoS attacks, others are also prominent, as is further discussed in Chapter 8. Ransomware, phishing, worms, and other assaults all become easier through Thing proliferation. The vast expansion of Things means more opportunities for attack, and the increasing number of serious attacks indicates that these unintended vulnerabilities are indeed being exploited.

On a network level, these vulnerabilities impact consumers, which is dangerous enough. However, with the expansion of Things into the industrial sector, the critical infrastructures of the United States and other nations are also in danger. As other nations besides the United States sponsor hackers, cyberspace, the new frontier of warfare, waits to be exploited through the needless vulnerabilities posed by Things today.

Beyond even these substantial security concerns, the ease of hacking into Things also raises stark privacy risks. Consider José and Miguel's home-security system. While providing some protection, the system would alternatively reveal the state of the home's occupancy. If hacked into, or viewed over Miguel's shoulder by a passerby, the system could ironically be used to facilitate a better plan for intrusion. Moreover, the system could also be used to spy on the inhabitants remotely. Such spying through home cameras happened in Houston, Texas, in 2016, with an unsecured camera in the room of an eight-year-old girl that had been installed by her parents.[283] Through an app called Live Camera Viewer, the footage was widely distributed for anyone watching to peruse. Poorly secured children's toys have been widely compromised across the globe. As just one recent example, in February 2017 it was

[282] Hautala, "Internet-connected homes open the door to hackers."

[283] Ptashkin, "Web Cam of Houston Children's Bedroom Hacked."

discovered that data from CloudPets—Internet-connected stuffed animals—was easily accessible.[284] Voice recordings of children and their parents along with the account information of eight hundred thousand users were available in a database that its designers failed to encrypt, while relying on default passwords.

The idea of regulation for Things, with systematic implementation of security by design, is moreover not a position reserved for Luddites. IBM, among others, has proposed it as a viable method to improving IoT cyberdefense which remains sadly lacking.[285] Privacy by Design should also be worked in more systematically while the opportunity is here.

The IoT certainly offers a great number of benefits. Nevertheless, its development must better account for its massive vulnerabilities. Threatening our infrastructure, economy, and personal privacy, the possible dangers are severe enough that they should be taken seriously. Yet, as the recent past has indicated, we clearly cannot just hope for an enlightened state; we must actively implement it. IoT proliferation should hence be paused until these concerns are addressed.

Big Data's Big Potential for Harm

While the revolutionary power of Big Data must be acknowledged, it also harbors a darker side. Critics who claim Big Data is useless are obviously mistaken, and it has seen great success and continues to produce vast benefits. However, as algorithm use mentioned in Chapter 4 illustrates, Big Data can also be used to reinforce the status quo and justify discrimination in a self-perpetuating cycle. Cathy O'Neil, for instance, has highlighted how the justice system, in looking for particular types of crime and using loaded measures of

---

[284] Hautala, "Smart toy flaws make hacking kids' info child's play."

[285] For instance, see Bonderud, "IoT Security: The Cold, Hard Truth About Digital Device Defense."

recidivism, merely reinforces stereotypes as to what "sorts" of people pose the highest risks to society.[286]

SPOTLIGHT: WHAT DOES YOUR SMART FRIDGE KNOW?

An ongoing court case, the James Bates murder trial hinges on smart-device data that was used by law enforcement to identify the main suspect's whereabouts. In this case, Amazon's Alexa passively collected data even while ostensibly turned off. Arkansas law enforcement sought access to Bates's Alexa files in addition to readings from his home energy meters and other smart devices. Initially, Amazon would only provide Bates's account information and purchase history, requiring law enforcement to meet the heightened burden for disclosure of audio material (as per the Electronic Communications Privacy Act from 1986, this includes showing that all other means of acquiring potentially incriminating information have been exhausted).

Though Bates voluntarily agreed to share the data with police, the court ruling may finally give an answer as to whether individuals have a reasonable expectation of privacy for data collected from smart devices. It may also reveal how much import is to be given to smart data (Paul, "How Smart Devices Could Violate Your Privacy"). This is a case bound to live awhile longer in the courts and appeals process. However, the smart devices themselves that flavor the case in such an interesting way remain immediately relevant. Are you aware of how much data your smart devices collect? Do you approve?

Typically, Big Data—owing to its volume, variety, and velocity—cannot simply be perused by the human eye. Millions of variables are collected at millions of different times before analysis. These data points, for instance, might include all the different readings from sensors around

[286] O'Neil, *Weapons of Math Destruction,* 84-100 discusses stereotype reinforcement in the justice system and offers a possible solution.

a hydroelectric plant, with data collected every fifteen minutes. With data incoming at high velocity, apparent trends deducible at an earlier time may no longer be valid in the present. For all of these reasons, Big Data analysis is best performed through machine learning, in which a computer program parses through the collected data. The computer program may be given particular parameters if the human analyst overseeing the project already has an idea in mind. However, just as well, the program can essentially be left to itself, using a series of statistical techniques to identify correlated parameters. From some basic data set and perhaps further human testing, the program can identify various algorithms that model the data and identify whatever outcomes may be desired. In turn, this can lead to a better design of new Things or improved functionality of current ones (see Chapter 8).

In itself, the process is not problematic. Critics who claim that Big Data projects cannot honor the privacy principle of Purpose Limitation,[287] whereby data is used only as the collectors indicate upfront, are mistaken. Mining Big Data does not mean either that the miners have no knowledge of what they seek or that users cannot know the intended purposes of researchers. Were researchers to simply reveal that they are looking for strong correlations between data elements—searching for "interesting" connections, for example—that would be very informative for the data subject. However, the issue arises when Big Data is applied under poor premises—such as the Google Flu Trends case[288]—or else when it is thought to be infallible.

O'Neil provocatively terms misapplications of Big Data "weapons of math destruction." While somewhat hyperbolized, the essence is true: analysis that fails to engage critically about flaws in the data set or that equates correlation with causation is bound to do more

---

[287] See, for instance, Eggimann and Tamò, "Taming the Beast: Big Data and the Role of Law."

[288] For more on the case, see Chapter 4 and Lazer and Kennedy, "What We Can Learn from the Epic Failure of Google Flu Trends."

harm than good (see Chapter 4). This applies to Thing data as well as any data in general.

The Dark Side of an Internet of Everything

Cisco and numerous other companies are interested in expanding the concept of the IoT into an Internet of Everything (IoE).[289] True cynics might claim the object is world domination, but more likely these companies do harbor hopes for a more deeply interconnected world. Yet, is the dream itself really desirable? On account of privacy and other concerns, there is strong reason to doubt jts worth.

A more interconnected world does not mean an absolute loss of privacy. Some protections can be established so that there remain separate public and private spheres. For instance, what one does at home, what one reads, and how one votes can still remain private affairs. However, with the public today even more public than before, and the domain of the private restricted, privacy would still be severely limited.

It is telling that from Sir Thomas More onward, nearly every single depiction of a utopian interconnected world presents significant danger lurking beneath a pleasant façade. Is this simply because conflict proves more compelling literature? Is it born from fear of the unknown, that all-too-familiar fear that motivates much xenophobia and induces us to keep things as they are, restricting innovation along the way?

Or is it perhaps because such a deeply interconnected world is fundamentally undesirable? Of course, how we determine desirability depends in part on how we evaluate privacy. If privacy is something integral to our flourishing, then we cannot let it perish. A wholly connected world, however, does preclude us from holding secrets. It makes public where we are and pressures us to don only a single face that conforms to the one worn by society as a whole. The proponent of such a world might suggest it is better to be honest and true. Yet, the result of an Internet of Everything is unlikely to be either. After all, do

[289] Techonomy, "Business and the Economy in an Internet + Everything World."

we not need to try on various perspectives and even widely divergent views before we can settle into something more comfortable?

Conversation is an opportunity for invention and reinvention. It is a time to explore and engage with different approaches. Different cultures do have different conventions. Some who prefer direct discourse might immediately jump into the business at hand. Others prefer to ease into business after developing a deeper relationship. Neither is inherently right or wrong; they are merely different. Depending upon our own background, we may have preferences for one over the other, but there are no objective criteria by which to decisively rank the conventions. Each offers different benefits and different sets of drawbacks. We are already in a world where limits for exploration are being set. Individuals are judged quickly, sometimes based on a single tweet of 140 characters, and sometimes for their entire life thereafter.

Public disapprobation is a social rule, the prerogative of any people. Those who say and do obviously offensive things perhaps ought to be treated with ridicule. Yet, the effect of a truly interconnected world is to undermine honesty to one's own mind and dam the personal streams of logic and emotion from flowing to their ends. Even with protections such as restrictions on data access and data storage life, the broader surveillance and tracking powers necessary to a more interconnected world would be considerable.

We already have massive communication capabilities that undoubtedly do benefit us on the whole: we can contact our loved ones and kindle friendships much more quickly now than ever before. We can message and video chat with people halfway around the globe with ease. Having greater communication capabilities, however, is significantly different from having all Things be interconnected. With devices vulnerable, a thermometer, energy meter, or pacemaker can be transformed into a terrorist threat. While the risk of any particular individual being targeted may be low due to sheer numbers, the risk of some device being targeted is extremely high. Even if we limit our use of Things to localized networks for defense purposes, there will remain a substantial risk. An even more interconnected world would further undermine personal privacy and freedom of thought. These are costs too high to bear, and as such the IoT must be checked more carefully.

In Conclusion

The IoT and the Big Data that both feeds and is fed by it offer myriad benefits and dangers that are technological, political, and social in nature. Whether the technologies are moving too fast or not fast enough, the IoT is here to stay. Just as José and Miguel grapple with leveraging the security benefits of their Thing with practical and other complications, so too will many of us have to grapple with our own networks and Things. What we opt for will impact the shape and sounds of our physical as well as mental worlds for years to come.

Further Exploration

Perhaps the first major popular use of Big Data, Michael Lewis's ***Moneyball*** follows Billy Beane circa 2002, the then-GM of the Oakland Athletics, as he sought to produce a winning team despite limited funds. His approach, alongside Paul DePodesta and a team of scouts, utilized analytic data related to baseball in order to evaluate which players could best assist the team. See also the 2011 film of the same name, adapted by Bennett Miller and Aaron Sorkin, starring Brad Pitt.

A compendium of further sources related to **smart cities** can be found at the **Future of Privacy Forum**'s "Smart Cities Resources" page: **https://fpf.org/smart-cities-resources**. The **Smart Cities Council** is an international effort to build smart cities embodying "livability, workability, and sustainability." More information is at **https://smartcitiescouncil.com/article/about-us-global**. News publication **ZDNet** has moreover dedicated a special tag to stories on "Smart Cities," accessible at **https://www.zdnet.com/topic/smart-cities**.

Privacy International has compiled several of the effects that Big Data has had or will have, focusing particularly on the social risks: **https://www.privacyinternational.org/privacy-101**.

A leader in identifying cyberattacks around the world, **Norse Corp** has received some criticism regarding its reputability.[290] Nonetheless, through its live map of cyberwar, it provides some insight into the proliferation of cyberattacks, many of which are triggered by Things: **http://www.norse-corp.com**.

[290] See, e.g., Ragan, "Norse Corp Disappears Shortly after CEO Is Asked to Step Down."

In March 2018, the UK government released a report entitled "Secure by Design: Improving the cyber security of consumer Internet of Things." This report builds on an earlier concept **"Privacy by Design"** (PbD), chiefly attributed to Ann Cavoukian, the Ontario Privacy Commissioner circa 1995. As the name suggests, PbD is a strategy to implement privacy measures, such as limiting data collection only to what is necessary, into the fundamental design of a device or service. More on the implementation of PbD in Canada and across the globe can be found in Cavoukian's 2011 White Paper: **http://www.ontla.on.ca/library/repository/mon/25008/312239.pdf**. Likewise, **"Secure by Design"** recommends building security into devices and services. It proposes a code of thirteen principles that chiefly assign responsibility to device and service providers, rather than customers. These include avoiding default passwords and minimizing exposed attack surfaces: **https://assets.publishing.service.gov.uk/government/uploads/system/uploads/attachment_data/file/686089/Secure_by_Design_Report_.pdf**.

Shodan provides users the ability to search for Internet-connected devices and learn of their basic specs, including security configurations. It mainly lists consumer devices but also includes some industrial ones. Some of these devices may be intended as stings to turn predators into prey, although the vast majority are apparently legitimate: **https://www.shodan.io**.

Chapter 8: Cloud Control and Information Security

Key Questions:

- What is the "cloud"?
- Who can access data stored in the cloud? Who controls it? And who owns it?
- How secure is data in a cloud? And what are major threats?
- What are best practices for cloud use today?
- How does international data law activate when data is in a cloud? Who has jurisdiction? Who would prosecute any cloud crimes?
- What are some other specific cybersecurity threats to which a cloud can serve as a solution or else activate an additional risk?

> Consider: Ava, a university student, is interested in purchasing a new cloud storage service. Caught in a recent spell of bad luck, she is eager to turn the page. Her tangoes with travail began at the public library a few months ago on a scorching April day, when she purchased a water bottle at a vending machine and left her laptop unattended in a study booth. The computer was gone when she returned, and though she had already transferred many of her files to an external hard drive, her latest few assignments, including several in progress, were lost. A few weeks ago, her replacement device, overworked from excessive tabulation, crashed. Only an older backup could be recovered from the hard drive. And a few days ago, Ava had another scare when a friend accidentally spilled apple juice on her phone.

Though Ava has tried to be diligent about transferring key files to an external hard drive—especially in light of these recent events—the time required for transfer is a hassle. Hence, she is now looking for other ways to back up her important files. A cloud storage plan from a prominent technology company seems like an ideal way to do this. One particularly attractive plan offers initial storage of several gigabytes, with an option to increase storage space for an additional monthly rate, if necessary.

Ava has also read news reports of prior data breaches at this company, including one two years ago. With an upcoming course on patents involving access to some confidential documents, Ava wants to avoid new misfortune with her files. Should Ava purchase this plan, seek the same service from a smaller firm, or avoid the cloud altogether? Are the dangers of data breach too great to overcome? Are there any better options available?

—Inspired by various cloud computing plans

Like *the Internet of Things,* like *algorithm*, and like *drone*, the term *cloud* remains something hazy, host to disparate interpretations. To obscure the picture further, the limits of the digital cloud are even more amorphous than its physical namesake. To the naked eye, water vapor and gas molecules determine the boundaries of the precipitative clouds that hover in the sky. Yet, in both popular usage as well as the parlance of various technical professionals, *cloud* simultaneously designates the Internet as a whole, particular segments of the web, and entirely separate dynamic processes. Though the cloud is intimately linked to the notion of an Internet (and particularly the Internet of Things), its domain is unclear. The first step to understanding the impact of the cloud will hence be to better sculpt its edges.

Catching the Cloud: Definitions and Scope

Frequently, *cloud* is combined with *computing.* A motley group of definitions includes the following:

(1) Singer and Friedman: "Cloud computing [is] the concept of delivering computing resources remotely over a network."[291]

(2) NIST: "Cloud computing is a model for enabling ubiquitous, convenient, on-demand network access to a shared pool of configurable computing resources (e.g., networks, servers, storage, applications, and services) that can be rapidly provisioned and released with minimal management effort or service provider interaction."[292]

---

[291] Singer and Friedman, *Cybersecurity and Cyberwar: What Everyone Needs to Know*, 106.

[292] Mell and Grance, "The NIST Definition of Cloud Computing."

(3) The World Privacy Forum: "*cloud computing* involves the sharing or storage by users of their own information on remote servers owned or operated by others and accessed through the Internet or other connections."[293]

(4) 2008 Pew Report: "['Cloud computing' is] an emerging architecture by which data and applications reside in cyberspace, allowing users to access them through any web-connected device."[294]

These descriptions share many similarities, alluding to or directly referencing the cloud's nearly ubiquitous accessibility via an online network and its location in virtual rather than physical space. Importantly, while "computing" is key to the cloud's functionality, this need not entail a consistently active process. As several of the descriptions above suggest, cloud computing may entirely consist of digital storage, the sort sought out by Ava.

Several other descriptions have expanded on the shorter definitions to highlight elements key to most, if not all, types of cloud computing. Jeffrey Rosenberg and Alfred Mateos, for instance, have offered five principles that they believe generally capture it:

A. Pooled computing resources available to any subscribing users
B. Virtualized computing resources to maximize hardware utilization
C. Elastic scaling up or down according to need
D. Automated creation of new virtual machines or deletion of existing ones

---

[293] Gellman, "Privacy in the Clouds: Risks to Privacy and Confidentiality from Cloud Computing," 4.

[294] Horrigan, "Use of Cloud Computing Applications and Services."

E. Resource usage billed only as used.[295]

Alternatively, Joe Weinman, author of the popular *Cloudonomics*, has defined the cloud as a service with five necessary elements: common infrastructure, location independence, online accessibility, utility pricing, and on-demand resources.[296] Altogether, these quite cleverly spell out "CLOUD." Meanwhile, another prominent categorization is XaaS: "X as a service." Most commonly, these feature SaaS (Software as a Service), PaaS (Platform as a Service), and IaaS (Infrastructure as a Service).

Articulated by specialists working intimately with the cloud, the vast majority of definitions have, perhaps naturally, stressed the cloud's benefits. Particularly of note are the cloud's ubiquitous accessibility, the diverse services available, and the ability of users to rapidly scale billing up or down to their needs.

Scalability has likely been a key reason for the cloud's recent growth. For instance, in 2011, after a White House report revealed that many agencies were using less than thirty percent of their server capacities,[297] federal agencies began seeking more efficient options. Another estimate has found that, from 2010 to 2014, the cloud computing market grew from $79 billion to $149 billion.[298] The market and the cloud's high functionality via storage, computational ability, and diverse products continue to expand. This is further highlighted by the varied leaders of cloud services today, which collectively provide motley services. The leaders include Salesforce.com/Force.com, Microsoft Azure, Amazon Web Services, the Rackspace Cloud, GoGrid, Facebook, Twitter, Google Search, Google Apps and App Engine, Apple iTunes

---

[295] Rosenberg and Mateos, *The Cloud at Your Service: The When, How, and Why of Enterprise Cloud Computing*.

[296] Weinman, *Cloudonomics: The Business Value of Cloud Computing*.

[297] Kundra, "Federal Cloud Computing Strategy."

[298] Transparency Market Research, "Cloud Computing Services Market"

and the App Store, and Wikipedia.[299]

Accessibility of services is another major benefit of the cloud, and also a key reason why it is intimately linked with Thing devices. For portable and multipurpose Things, computations and other services are most effectively provided remotely: they are best supported through cloud services. As Chapter 7 indicated, Thing connectivity comes with its own set of benefits and hazards. For instance, while the Internet of Things may boost efficiency and increase the prospects for innovation, it also introduces new security vulnerabilities.

In this chapter, we will build on the promise and perils of interconnectivity in the IoT to consider other aspects of cloud use. Particularly, we will consider technical complications resulting from data storage and ownership, various species of cybersecurity risks, and the implications of transnational operations. For each of Ava and others considering investment in the cloud, these competing considerations will likely have different levels of importance.

A Case For Increased Regulation of the Cloud

The emergence of nearly any technology, shiny and new, inspires vocal opposition. Sometimes this resistance manifests violently, as with the Luddites who opposed industrial machines in the late 1800s.[300] Other times, opposition is reflected in popular derision. Such was the initial fate of umbrellas, criticized as a fashion faux pas in more gender-sensitive times. In other instances, early critics may seize upon a genuine defect in a new product, but write it off all too quickly. This was the case with the personal laptop, a still-bulky beast in the mid-1980s. And in still other

[299] Weinman, *Cloudonomics: The Business Value of Cloud Computing*.

[300] As discussed in Conniff, "What the Luddites Really Fought Against," the original Luddites' opposition to technology may be more nuanced than is generally acknowledged today. Nevertheless, their opposition to industrial machinery was certainly potent.

situations, as with early "computerphobia," a particularly radical technology may trigger all sorts of anxieties—from whispers of the occult to fears of destroying something expensive or fears of proving one's stupidity.[301]

Undoubtedly, some of the hysteria surrounding new technologies is ill-founded, a knee-jerk reaction to change. Nevertheless, when it comes to cloud computing, there are legitimate reasons to believe that it presently poses a serious threat to individuals and society as a whole. These include deep-rooted cybersecurity dangers that it creates—dangers which have already been exploited. Additionally, present environment problems intensify the clouds' risks with the present disrespect of personal data and jurisdictional confusion. In order to adequately deal with these concerns, regardless of short-term benefits the cloud might provide users such as Ava, it must be regulated more strictly and severely restricted in scope.

Deadly Databases

In the mid-1960s, the federal government strongly considered the implementation of a national database to consolidate personal information. This database would include each citizen's electronic birth certificate, proof of citizenship, school records, draft registration, military service records, tax records, Social Security benefits, and death certificates. Much of this data was already collected by separate agencies (such as the Social Security Administration and Bureau of Labor Statistics), but consolidation was a new endeavor.[302] As the program's advocates hoped, through the proposed database, the bureaucracy could

[301] For more on umbrellas' adoption struggles, see Waters, "The Public Shaming of England's First Umbrella User." Sandberg-Diment, "The Executive Computer" presents an early eulogy for laptops, and Lafrance, "When People Feared Computers" discusses various manifestations of "computerphobia."

[302] Garfinkel, *Database Nation: The Death of Privacy in the 21st Century*.

be streamlined and efficiency expanded, with further benefits such as identity protection.

Following the publication of a *New York Times* article critical of the proposal,[303] Congress began hearings that ultimately spelled doom for the program. As the article highlighted, such a database would concentrate power, and as another critic noted at the time, data concentration could be "catastrophically dangerous."[304] Today we know how far a little data will go (see, for example, Chapters 2 and 3). And a centralized database, would contain more than just a little, whilst providing a single, convenient target of value. The hope for the proposed national database was always expansion, and once the infrastructure stood, growth of the content within could accelerate. Voter data and financial records, for instance, were proposed as possible additions. In turn, not only would the database contain personal biographical information for lookup, but it could also be used as the basis for validation, such as of tax payments and citizenship. Of course, this could also be used for malice.

Just a few years after the database deliberations, Watergate illustrated paranoia's legitimacy. The Watergate plumbers, at then-President Nixon's request, bugged the Democratic Party's headquarters so as to learn and counterattack their election strategy, potentially also gathering information for blackmail. Though it was unlikely that the data gathered contributed in any meaningful way to Nixon's landslide victory, the incident certainly demonstrated how far a motivated politician could go to subvert the democratic process for personal gain. With a comprehensive national database in his arsenal, Nixon or future leaders could well have subverted democracy. For instance, they could "accidentally" erase citizenship records of those in an opposition party, or denote personal enemies as criminals, ineligible to vote. It would take

303 Packard, "Don't Tell It to the Computer."

304 Ibid.

just one motivated individual to erode the pillars of justice. Thankfully, the comprehensive national database did not then emerge.

Nevertheless, aggregation of data has again become a keen focus. Though not quite on the scale of the previously proposed directory, personal information is compiled today by data brokers and law enforcement, as discussed in Chapters 3 and 6. Meanwhile, as addressed in Chapter 9, genetic data is beginning to be stored in repositories by private corporations. Cloud computing, particularly storage, enables these and other dangerous databases. It also does so in a way that exacerbates the vulnerable qualities.

Inherently, several of the cloud's desirable attributes, also increase vulnerability. In particular, the cloud's easy access via the web and via assorted Thing devices, activates a new weakness: there are more entry points to be exploited. Furthermore, use of the cloud requires transport of data from a local machine. Like banks receiving gold in armored trucks, with security officers standing by whilst carrying fearsome artillery, these data transfers may be well-secured according to industry protocols. However, the mere fact of transfer opens a new avenue for attack that committed individuals will seek to exploit. And agents certainly have sought to exploit these vulnerabilities. Two separate compilations of publicly disclosed data breaches are available online, one by the nonprofit Privacy Rights Clearinghouse and another by Maryland security firm Gemalto.[305] Gemalto's Breach Level Index has noted more than nine billion data records lost or stolen since 2013. Tracking publicly disclosed data breaches around the world since 2005, the Privacy Rights Clearinghouse has identified an overall increasing trend in the number of global attacks, along with colossal increase in records affected. For instance, 2005 saw 136 total breaches worldwide, affecting around fifty-five million records. 2015 saw about 319 million records affected, while 2016 featured 782 attacks affecting an astonishing

[305] See the "Further Exploration" section of this chapter for links to these, and the cited statistics.

four billion records. As of June 2018, the numbers for 2017 were at 540 noted attacks and more than two billion affected; though a dip from the year prior, the trend upwards is unmistakable and incidentally, since it may well take American companies an average of two hundred days to identify breaches, we can expect the 2017 numbers to increase further.[306]

One particular breach, discovered in July 2017 and reported two months later, was only as impactful because data had been consolidated. Equifax is one of the Big Three credit reporting agencies, and hence compiles personal data for nearly every individual with a credit history. This information includes social security numbers, birth dates, addresses, and registered license plate numbers. Through a hack perpetrated over an estimated two months—and one that could have been weakened substantially had a software patch been applied when recommended three months prior—the private information of more than 143 million individuals was exposed. Specifically, more than 200,000 US credit numbers and expiration dates were collected, along with the government identification records from more than 182,000 customers. The full extent of the breach's fallout remains to be seen, as more individuals' data may have been collected without a trace.[307]

Through the prolific number of entry points connected to the cloud, the possibility and impact of traditional cyberattacks—including distributed denial of service (DDoS), ransomware, phishing, and worms—also increase in severity. In DDoS attacks, as mentioned in Chapter 7, a number of Things are coordinated to overthrow a network. The attack's effectiveness is only possible through the volume of interconnected devices operating on the cloud, and, including the

---

[306] For instance, see the Ponemon Institute's 2018 "Cost of a Data Breach" report, linked in this chapter's "Further Exploration" section.

[307] For a brief summary of the incident see Knight, "Equifax reveals full extent of last year's massive data breach." See also the "Further Exploration" section of this chapter for more detailed information, including the company's official accounting of the incident.

October 2016 Dyn attack, DDoS impacts have already been immense.[308]

Through ransomware, a hacker obtains access to some service—such as a database—and locks out the proprietor until a financial ransom is paid. Of course, there is no guarantee that once the money is paid the data will be released, but a noble principle to "not negotiate with terrorists" will certainly not retrieve stolen data nor will it restore clients' confidence in the company in the event of the attack's discovery. One large ransomware attack occurred in early May 2017, when a vulnerability in Windows was exploited by hackers (suspected to be North Korean agents) at hospitals, insurance companies, and many other facilities. All told, three hundred thousand machines were affected by the WannaCry attack, and more than $146,000 was paid by those affected.[309] Less than two months later, another ransomware attack struck worldwide, this one originating from Ukraine.[310] Once again, consolidated data became a vulnerability, exploited. A plausible solution would be reliable backup storage; however, due to its value, health data could not just be easily passed around, and certainly not readily propagated to the cloud. This

[308] Already in 2014, security firms were noting the increased pace of DDoS attacks for extortion and costly service outages, as described in Apps, "DDoS Cyber Attacks Get Bigger, Smarter, More Damaging." Massive attacks sprang from both within the US as well as abroad. Following Dyn attack, forensics turned out inconclusive, with some suspecting state-sponsored Russian hackers, others laying the blame on perky youths, as described in Lomas, "Dyn DNS DDoS likely the work of script kiddies, says Flashpoint." While the attackers may remain publicly unknown, the continued danger of denial of service attacks is not. As considered in Lewis, "The DDoS Attack Against Dyn One Year Later," new versions of the Mirai malware which converted users' Things into zombied weapons have already been developed, and possibly deployed.

[309] Chappell and Penman, "Ransomware Attacks Ravage Computer Networks in Dozens of Countries" and Collins, "Victims of the WannaCry Ransomware Attacks Have Stopped Paying Up."

[310] Goldgeir, "Ukraine Cyberattack Wasn't Really About Ransom, Security Experts Say."

would necessitate local storage, making redundant cloud computing's ability to do the same, at least in this context.

Worms are another common mode of attack, such as the Stuxnet bug that infiltrated Iranian nuclear systems. Self-replicating, worms can infiltrate a system and lie dormant until a designated time for assault. Often, a worm is introduced after a phishing attack, where a cyberattacker probes an unsuspecting user for information (such as access codes) under the pretense of authority or impending catastrophe.[311] After acquiring access information, attackers can send in a worm that slowly infects more and more of the target. It may begin with capture of a single user's data, but buoyed by the cloud's interconnectivity, this entry point can allow access to a connected database in its entirety.[312]

Thus far, defenders have largely been able to quarantine and quell cyberattacks. However, attacks such as the Dyn assault, WannaCry, and the Equifax breach have already affected millions, and their potential for destruction moving forward cannot be underestimated. A robust solution is in order, but the cloud only appears to exacerbate current problems.[313]

---

[311] For instance, emails warning that "Your account has been suspended" are common. These messages may appear to come from a legitimate source using a trusted company's logo and colors, and sometimes even an e-mail address that closely mimics a corporate design. Yet, the subsequent links that ask you to re-enter identifying information such as credit card details are only pretenses to phish the information from you. Appleby, "What is Phishing" offers several examples of real phish.

[312] The cyberattacks discussed here are described in broad terms. Other sorts of attacks also exist, along with subspecies of those addressed. More information on such cyberattacks can be found in the "Further Exploration" portion of this chapter.

[313] There are many compilations of major cyberattacks in recent years. One compilation with links to more information on the Equifax breach as well as others can be found at Larson, "The hacks that left us exposed in 2017."

Another danger of cloud storage is also tied to its perceived value. Remotely accessible, data stored in the cloud is immune from local possible catastrophes. As such, it is also more difficult to ensure that data is deleted when appropriate. Even on personal computers, file deletion is not as easy as a tap of the delete key might make it appear. The delete key may remove an icon from the desktop or even the user interface in the program listing, but the file remains in a purgatorial recycle bin, awaiting ultimate destruction. And even if deleted from there, other copies of the file may persist on the computer through cached files intended for restoration in the event of a sudden shutdown. These backups could well be useful to individuals such as Ava, who have suffered from sudden computer shutdowns. However, they also allow for thieves to find personal data on files such as bank statements, credit reports, and medical reports that were apparently destroyed.

For the cloud, tracing the path of data for deletion may be even more complex. As with existing cybersecurity standards, this can be rectified through better-conceived protocols and their enforcement, whereby deletion abides by a standard process. However, the infinite nature of the cloud, as well as the secrecy that cloud owners justifiably maintain about certain structural elements, mean that the data can roam. This roaming might be the cause of backup procedures used by corporations in the event of compromise. It might also be the result of law enforcement demands. As the Snowden leaks revealed, data from telecom companies including Verizon and AT&T, as well as market dominators Google and Facebook, has been seized for analysis in the past, and still can be. Perhaps creating an illicit version of the database rejected in the sixties, surveillance agencies empowered by FISA Amendments, the USA Patriot Act, and other legislation discussed in Chapter 6 can today dive into consumer data.[314]

Finally, one further negative consequence can arise unintentionally. The cloud's ease of access and ease of use invite users

---

[314] MacAskill, "Yahoo Files Lawsuit against NSA Over User Data Requests."

to rely and in time depend on it. Mistakes ingrained within, however, can then end up haunting individuals and families for years. This has occurred in the past with regard to taxes and credit scores. For instance, the IRS has failed to receive tax checks because of mail delivery mistakes, with a post office erroneously rerouting an envelope. Individuals with the same name have also been mistaken for one another, resulting in the wrong man being designated as a tax evader.[315] Though the 1970 Fair Credit Reporting Act did grant greater visibility into credit records, in part to alleviate such mistakes, this was not an absolute transparency, nor did it promise accuracy. For one, if individuals disputed any claim on their record such as one of tax evasion, even if producing indisputable evidence to illustrate the error, the evasion note would not be expunged. Rather, in a practice that persists, an addendum would be added to the affected individual's file, explaining the error. Such a note could be trusted to suffice if consistently acknowledged, but the reality is that individuals cannot see all the significant data in their files. In the past, credit companies and banks have produced customer credit scores from one to ten to represent creditworthiness. Consumers have no insight into these values, and though credit-assessing companies insist these numbers do not affect credit decisions, the truth of this is unclear. The real-life cases of various individuals, innocent but tainted with accidental black marks, according to which they were denied new credit cards and loans, suggest the immense power of these invisible numbers.

Before the new millennium, it was standard practice that credit records last only seven years before being updated.[316] As such, after seven years (and ten in the case of bankruptcy), mistaken notes would be removed. Since the Internet has come of age, however, and cloud storage has allowed for a library of the undead, these records can be

[315] These and other unintentional but consequential mistakes are discussed in Garfinkel, *Database Nation: The Death of Privacy in the 21st Century*, 13-69.

[316] Ibid.

preserved for much longer. Mistakes, hence, can become impactful data as never before.

Cyberattacks affecting the cloud can hence range from minor inconveniences to calamitous shutdowns. Data may be stolen for its own value or for the information (e.g., bank account numbers) to which it links. Alternatively, access to important data troves may be obstructed, such as through ransomware or denial of service attacks. While cybersecurity risks affect all aspects of the Internet of Things, they are especially potent for centralized clouds. Attacks on these central spaces, via malware and other means, persist. Ultimately, the power of cloud computing means that the target is likely to be tantalizing for deranged agents. And this is distinct from the fear that the current infrastructure has already been compromised.

The recent proliferation of attacks highlights the serious vulnerability of clouds and connected devices today. Some of the problems may have relatively easy fixes—even if, shockingly, they have not yet been implemented—but others are deeply rooted. Hence, until a broader solution is incorporated, use of the cloud must be curtailed: it should not be allowed to irresponsibly spread further.

The Status Quo: Blatant Theft of Personal Data

Above these deep-seeded dangers, the current ecosystem bears a heavy weight upon any personal data collected and stored by the cloud. Essentially stolen by corporations, personal data's disrespect indicates that necessary change is unlikely to come if the cloud is left to its own devices.

In the United States today, companies sell consumer data for sizable profits without individuals' clear consent. Genetic profiling companies do this, as discussed in Chapter 9, with portions of personal genomes patented for profit. Moreover, even in other sectors, companies have sold personal data, such as when retailers RadioShack and Sports Authority each sold customer information as an asset when facing bankruptcy.[317] Such sale is presently legal, provided that there is a provision in these companies' privacy policies stating that personal data may be transferred. Certainly, individuals are responsible for vetting what they sign away; ideally, they would only do business with those corporations aligned with their interests. However, there is not a serious choice available today. For one, corporate privacy policies are highly obtuse, and difficult for ordinary readers to comprehend.[318] Moreover, "transferal" provisions are commonplace, which is what substantiates the data broker market discussed in Chapter 3. Hence, even if customers parsed companies' complicated legalese, few alternatives would actually be available. Additionally, despite some recent furor over prominent data breaches and cyberattacks, it appears that sale of personal data is expanding within the United States.[319]

To preserve the choice undergirding privacy, a fundamental shift is in order. From the government to the companies incorporated, our society stands by and for the people. Big companies should not be profiting at the expense of a society's individuals, and yet that is precisely what is happening now with personal data. As addressed in Chapter 2, personal data cannot be sold away, and it certainly cannot be transferred to some faceless or multi-faced corporation. Rather, it always belongs to the individual. This fundamental idea is codified in the European Union General Data Protection Regulation (GDPR), which entered effect in

---

[317] Ciccatelli, "Selling Customer Data in Bankruptcy."

[318] Carnegie Mellon researchers in 2008 estimated that it would take approximately 244 hours per year for an average web user at the time to read all the privacy policies for services in which he or she was engaged: McDonald and Cranor, "The Cost of Reading Privacy Policies." Our dependence on varied services has only increased since then.

[319] On April 3, 2017, President Trump signed into law S.J. Res. 34, which enabled Internet Service Providers (ISPs) to sell consumer data, negating FCC regulations to the contrary from late 2016. Some more on the dangerous implications of this bill can be found in Chapter 2 as well as at Snider, "ISPs can now collect and sell your data: What to know about Internet privacy rules."

May 2018. Based upon prior privacy frameworks such as the FIPs from the 1970s and the OECD guidelines, the GDPR dedicates one of its eleven chapters exclusively to "Rights of the data subject." In Articles 12 and 13, the document requires clear communication by companies that collect personal data with regard to how the data is used, and permits individuals to withdraw consent at any time. With these and other provisions, the GDPR empowers individuals to control their personal data.[320]

SPOTLIGHT: WHO CAN CONTROL THE CLOUD?

The National Institute of Standards and Technology (NIST) is an executive agency originally founded in 1901 as the National Bureau of Standards. It had evolved from a constitutional directive to standardize "currency, weights, and measures" which served as the basis for economic and scientific inquiry in the young United States. The agency began to expand its scope to establish standards for government as well as consumer goods, ranging from chemical standards to forensics analysis guidelines to voting-machine parameters (NIST, "History"). NIST also sets standards for cryptography and information technology systems, including the cloud. Its numerous research endeavors have included missile and systems design. In its capacity as a federal agency, NIST frequently seeks out public comment before publishing reports on topics such as Bluetooth security (July 26, 2017), security of "Thing" networks (July 28, 2016), and US cybersecurity as a whole (forthcoming: NIST, "Cybersecurity Framework"). Unlike the FTC, NIST is non-regulatory, meaning that it does not enforce its guidelines. Nevertheless, these parameters still suggest best practices that can serve as guidelines to those in industry. The agency's publications can be located at the "Publications" tab of its website: https://www.nist.gov/publications.

[320] For more on the EU GDPR, see the "Further Exploration" section of this chapter.

This EU framework, which empowers individuals, should spread globally. Yet, these changes have been slow to be adopted in the United States and elsewhere. Hence, one alternative solution of the data-ownership inequity follows from the model of cloud computing to empower individuals. A micropayment, as the name suggests, is a very small payment in the ballpark of cents or fractions of cents. These payments would allow users to engage in a pay-per-use model, just like the one that cloud computing resources employ. Through micropayments, Internet content could be priced at nonzero-but-minimal rates. This would overturn the traditional landscape where content is either free or else available at a standard subscription rate.

On traditional fiduciary apps, tiny payment amounts still are not recognized, but with the growth of PayPal as well as Blockchain's Bitcoin (see Chapter 5), such payments have at least become possible. As far back as 1997,[321] proponents expected the market of micropayments to kick off, and yet today, their use remains limited to a few in-game purchases in smartphone apps.[322] Of course, it is not too late yet for these payments to expand. While there remain some technical limitations, the biggest hindrance seems to simply be inertia.

Yet, as various enthusiasts have proposed, micropayments could serve to make personal data's wild appropriation profitable for consumers until a more systematic solution is in place. As the idea goes, corporations would provide micropayment compensation to individuals whenever their data is used.[323] Each use might be differently priced depending on the importance of the datum; alternatively, it could be standardized, perhaps to a tenth of a cent or so. When added up over time, such micropayments could allow users to acquire a supplemental

[321] Negroponte and Pawson, "Pay Whom Per What When, Part 1."

[322] Keller, D., "Global Right to Be Forgotten Delisting: Why CNIL Is Wrong."

[323] See, for instance, Worden, "Micropayments and the Future of the Web" and Kaufman, et al., "Micropayments: A Viable Business Model?"

income in the range of fifty dollars or more every few months. Personal data would still have to be de-identified as much as possible for maximal protection, and other security measures would still be in order, but this technique would allow users to receive recognition and compensation, rather than being the victims of highway robbery. As such, individuals could maintain their stamp of approval and even prevent undesired sales.

Whether or not this particular monetized approach to personal data is adopted, the debate over how to appropriately value personal data needs to move forward, and it needs to advance now. Just as with a centralized database, if corporate data ownership is legitimized, the system will be much more difficult to overturn. Moreover, the difficulty of tracing data flow in the cloud further means that once personal data is gathered, the technical difficulty of deletion will multiply while the incentive for implementing the program will disappear. Through credit scores and other means, data brokers and tech companies are compiling opaque dossiers that may be incorrect and yet will impact our livelihoods. Control over personal data must return to the people.

Fragmented Regulations and Uncertain Jurisdiction

In conjunction with the mistreatment of personal data, the uncertainty of cloud governance enhances the cloud's overhanging dangers further.

Some regulations are in place, notably the EU GDPR in effect June 2018. The GDPR does provide a relatively robust framework as addressed previously. Moreover, it shifts jurisdictional focus from the location of a corporation's headquarters to the location of its customers' residence: if a company has data subjects who are inside the European Union, then at least for those customers, the company will have to abide by the prescribed Regulations, even if based out of San Francisco. Failure to do so risks fines greater than $23 million.[324] This is a particularly important designation for cloud computing. Due to the cloud's virtual nature and easy accessibility, cloud services can be—and frequently are—accessed from outside the nation in which the cloud service provider is incorporated. Hence, the old legislative regime which focuses on traditional national borders is inapplicable to this novel technology, and increasingly the modern economy.

With the cloud providing transnational services, it needs to be governed by transnational laws. Otherwise, there will be what there is today: mass confusion. Regulatory confusion is problematic for all involved. For corporations, it means unwitting missteps that risk substantial fines. For government regulators, it means uncertainty on how to proceed and codification of inconsistencies that impede future legislation. For individuals, confusion moreover means that wily corporations can shield their profiteering off customers' data, and continue to abuse it.

In March 2018, the United States did take one step towards jurisdictional clarification in passing the CLOUD Act. This act arose as a response to *Microsoft v. United States*, a case in which Microsoft claimed that American law enforcement did not have the authority to seize electronic communication records "stored" at the company's Dublin headquarters. Through the CLOUD Act, vociferously supported by the tech industry due to the clarity afforded, the US government is clearly authorized to access electronic communications even if stored abroad. Moreover, the bill establishes a new streamlined process for government requests of information. Additionally, the law requires American communications-service providers to cooperate with governments partnered with the United States via executive agreements. While addressing some problematic elements of confusion, this act appears to

[324] For more on the significance of this change, see this chapter's "Further Exploration." Specifically, Article 3 of the GDPR addresses the Regulation's broad scope, while the "GDPR Key Changes" page of the crowdsource resource linked points to this provision's novelty.

actually steal power further away from individuals, depositing it in the executive branch's hands. As several human rights groups have noted, a precise process of oversight for law enforcement requests remains undefined, meaning that other forms of intentional or accidental abuse may well activate as data propagates.[325]

Within the United States, more comprehensive regulations are in order, to not only clarify government's obligations towards seized personal data, but also to govern corporate treatment of personal data. Importantly, given the transnational nature of services possible through the cloud, regulations will also have to have transnational power. The cooperation proposed through the CLOUD Act, where executive agreements between the United States and other nations enter the force of law, is one solution that might be viable over time. Likewise, the EU-US and Swiss-US Privacy Shields, designed through partnerships with the United States Department of Commerce, provide some consistency for companies engaged in transatlantic commerce, and their customers.[326]

While international agreements offer one response, it is important to note that such agreements are effectively voluntary. A nation governs over its own citizens, but it cannot, on its own, enforce them elsewhere. Thus, there is the concept of asylum, where persecuted individuals can take refuge in one nation to avoid the reach of another. Treaties, which can be signed by legislative bodies, provide a means to enter government co-operations into law. However, in the United States, the executive is acknowledged as having "plenary" powers when it

---

[325] For more on corporate support for the CLOUD Act, see Hatmaker, "Consolidated Appropriations Act of 2018." Selected criticism of the the bill is at Guliani and Shah, "The CLOUD Act Doesn't Help Privacy and Human Rights" while Swire and Daskal, "What the CLOUD Act means for privacy pros" sketches in more detail the revised system by which law enforcement can now solicit communications data. The full text of the CLOUD Act is linked in the "Further Exploration" section of this chapter.

[326] See the "Further Exploration" section of this chapter for more information on these agreements.

comes to foreign affairs. This means that agreements by one president may be annulled by the next or vice versa, casting international agreements into a state of instability. Across the world, the United Nations, for all its intended glory, requires unanimity to engage in compelling action, and the fractured state of world blocs today means such unanimity is unlikely. For instance, a world power like China that openly sanctions state control of media is unlikely to support consumer protections; meanwhile, business interests in the United States will remain a lobbying force to be reckoned with. EU nations unhappy with a giant corporation could perhaps threaten fines or exile of any offending companies, but the tech companies are not powerless themselves, by virtue of the valuable services they provide, presently for free. In a battle between a tech giant and a nation-state, it is increasingly less clear that the nation-state will prevail. Hence, in order to preclude governance by the corporation, now is the time to pause and consider how the cloud world ought to develop to best serve us and what regulations will allow these corporations to do so. Thus, at their core, these agreements will remain somewhat tenuous; some degree of uncertainty will be inevitable.

Chapter 5 examined one effect of transnational confusion, whereby international search engines may have to modify search results in particular countries but not others. When it comes to uncertainty of cloud regulations, the source of friction has even greater impact and it runs deep into the very systems of governance that preside today. This does not mean that a happy and sustainable solution cannot still be found—the CLOUD Act, for instance, welcomes increased cooperation. However, given the deep dangers with the current state of play where personal data is mined for profit, given the cybersecurity vulnerabilities of centralized cloud computing, given the opacity propagated by current confusion, and given the continued growth of cloud computing to support even our critical infrastructures, we need to pause. We need to shape the cloud to serve us, instead of allowing it to balloon out. Otherwise, we risk passing a point of no return.

A Case For Continued Expansion of the Cloud

The cloud's greatest contributions are simple enough to express: efficiency and interconnectivity. These are valuable in themselves, and moreover, as with the Internet of Things and drones, these benefits expand into still others. For instance, they lead to an equality of opportunity that will benefit new businesses and empower a more robust inclusivity. Moreover, the ingrained characteristics of the cloud will feed efforts to bolster cybersecurity. In terms of the ecosystem, critics should not overreact; the current economic model in fact promotes consumer choice and autonomy. Additionally, while there undoubtedly is some inconsistency regarding cloud governance today, developed legislation indicates that progress is occurring. We should trust that appropriate law will develop as it has in the past, and we should not inhibit growth of a terrifically beneficial technology—a technology that is bound to produce new innovations and perhaps even solutions.

Ultimately, the best way to address present deficiencies will be to minimize regulations for the sake of growth and innovation. In turn, this will foster new solutions, solutions that benefit individuals such as Ava, businesses of all sizes from Maggie's Phishing With You to established players, and government agencies. For its bounteous boons, the cloud is understandably becoming integral to our economy and society; by fostering its continued growth we will in turn foster a brighter and more equitable future.

Expanding Efficiency Via the Cloud

Several of the definitions at this chapter's start highlighted fundamental benefits of the cloud. To reiterate several of these, the cloud allows users to scale their resource usage and costs at a moment's notice. In turn, this provides increased cost efficiency from prior models, such as local storage or a university department's wholesale purchase of an advanced computing machine. The remote and virtual nature of cloud resources further entails easier access than would otherwise be possible. Data backups, for instance, can be performed regularly and without the use of cumbersome wires or external hard drives. The easier uploads of data made possible by the cloud, moreover can serve to increase transparency and democratize analysis so that data sets can be viewed by a broader audience than might otherwise be possible, and so that meaningful conclusions can be drawn more swiftly. This has already seen fruition with data compiled by the United States Agency for International Development and environmental data that is being stored by the National Oceanic and Atmospheric Association, while several genomics projects discussed in Chapter 9 have a similar aim.[327]

By increasing time and cost efficiencies, the cloud also presents an equality of opportunity. As of the first quarter of 2018, the leaders in cloud infrastructure have been Amazon (33% of the market), Microsoft (13%), IBM (9%), Google (6%), and Alibaba (4%).[328] Each of these has reduced their prices in recent years, while also continuing to introduce new packages.[329] Cost-saving is valuable for any party, but it pays dividends especially for those on the rise. And in the case of the cloud, it has long serviced those up and coming: Amazon's AWS, for instance, was weaned upon startups, and it continues to retain the largest number, even as competitors are swooping in.[330] A chief reason for startups'

---

[327] See, Hayes, "How the Cloud Improves Transparency at the State Department, USAID and NOAA," and the "Further Exploration" section of this chapter for some more on these cloud-reliant efforts by the US government.

[328] Synergy, "Cloud Growth Rate Increased Again in Q1; Amazon Maintains Market Share Dominance."

[329] For instance, see Barrett, "Amazon's New Unlimited Cloud Storage is Absurdly Cheap," Lardinois, "Google's Cloud Platform gets a cheaper, lower-performance networking tier," and Hussain, "Cloud storage now more affordable: Announcing general availability of Azure Archive Storage," which chart how these providers have each continued to lower prices, while also introducing new packages that might suit customers' needs more closely.

[330] For more on modern startups' reliance on the cloud, see, for instance, Darrow, "Microsoft and Google Give Startups Options to Amazon's Cloud."

reliance on the cloud is clear: cloud services are extremely affordable and provide operational assets close to, if not on par with, those of large, established companies.[331] Via Platform or Software as a Service, companies can additionally deploy powerful analytical tools at affordable prices. By reducing companies' operational costs, the cloud is hence increasing the opportunities available to individuals and companies to innovate and thrive.

Among other extended fruits of efficiency is the fact that the cloud allows users to integrate solutions. Through the standard platforms available, companies can store and share data in sync with one another. This will have value for government projects, of course, such as the NOAA and USAID ones mentioned above. Moreover, for nearly any sector, data stored according to the same parameters reduces time and costs associated with synthesis.

There are bounteous other ways in which the cloud will benefit users via efficiency, some of which are addressed further through the references cited at this chapter's start. Still another benefit of the cloud is its value for cybersecurity.

Boosting Cybersecurity Via the Cloud

The ubiquitous accessibility of the cloud may raise some new vectors for attack. Nevertheless, history and the environment indicate that these vulnerabilities are not ones that should lead us to panic. Moreover, in numerous ways, the cloud actually serves to bolster cybersecurity.

First, the struggle between assailants and defenders is long-standing. In the wild, predators chase prey and those of the hunted who survive develop diverse defense mechanisms. These include clownfish who take shelter from bigger fish in the electric tentacles of anemones, antelope who are capable of zooming by at high speeds while changing

[331] For a shortlist of still other benefits of the cloud, specific to startups, see Jain, "Cloud communication helping startups grow, creating a win-win scenario."

direction with ease, chameleons that camouflage, mostly vegetarian snakes that mimic venomous kin, cacti that develop spines, and crabs that can autotomize a limb when under duress. In human societies as well, local and global conflicts persist, from ancient tribes, through castle walls, to cold wars, and now cyberwars. Regardless of the platform, attacks are hence inevitable. And as prominent cyberattacks indicated even in the web's youth, before the cloud had truly taken off, attacks are certainly inevitable in cyberspace.

Moreover, the true test of a system's viability is defense, an area in which the cloud soars. Consider, for one, the most prominent attacks that have been perpetrated. From their 2017 Data Breach Investigations Report, Verizon compiled a study of more than two thousand breaches from the years prior. The second-largest bloc of perpetrators (25%) consisted of "Internal actors." While organized criminal groups were responsible for around 50% of attacks, only 18% were associated with state-affiliated actors—those who would typically be associated with cyberwar. By another estimate—through a study conducted by security firm Radware in 2017, 20% of attacks were attributed to angry users, 27% were attributed to insider threats, and, again, less than a quarter to cyberwar. To some degree, of course, these studies will not be wholly accurate: breaches may not be discovered for months or even years, and some small breaches may never be widely reported upon. Moreover, the scale of these attacks is undeniable—one estimate of damage to the global economy due to 2016 cyberattacks stands at $450 billion.[332] Nevertheless, the percentages indicate that many of the attacks have not been the result of attackers exploiting some inherent flaw in data storage, cloud, or IoT technologies. Rather, attacks have primarily sprung from knowable and temperamental sources, ones which can be addressed and hence remedied in the future.

---

[332] The numbers cited above can be found at the Verizon Data Breach Investigations Report linked in this chapter's "Further Exploration" section; for non-subscribers, a summary of the details referenced here can also be found at DesJardins, "Why Hackers Hack: Motives Behind Cyberattacks."

The preventability of massive recent attacks is certainly clear. Consider again the 2016 Dyn DDoS attack described in Chapter 7. The largest such attack to-date, it was perpetrated through a network of ordinary Thing devices, including home cameras and DVR players. A subsequent denial of service attack in mid-2017 exploited similar vulnerabilities.[333] One major means by which hackers were able to commandeer Things was through insufficient security. Particularly, device owners either did not update the device passwords from factory presets, or were unable to do so.[334] Some manufacturers had designed their devices for easy use by consumers, without considering the ease with which consumers could modify login information; this was the case, for instance, with Hangzhou Xiongmai's cheap but readily accessible web cameras, which the company recalled following the attack.[335] To some extent, then, manufacturers were responsible.

Via inadequate security measures, customers were the other main vector for attack. Though the means were available, they failed to implement unique passwords. A number of simple solutions are viable, many of which can be implemented from the consumers' side. Fundamentally, consumers can and should personalize their devices. This means changing preset factory logins, such as username "admin" and password "123456," to user-specific credentials. Moreover, it means customers should customize device settings to best serve their needs. For instance, a phone app like Avian might be set by default to collect a user's location information. By managing one's device, a consumer can

[333] This latter, the BrickerBot attack, was designed to exploit similar vulnerabilities as the Mirai malware, responsible for the attack on Dyn. Some more information on this assault, compiled by security firm Radware, can be found at Geenens, "BrickerBot only attacks compromised devices."

[334] See, for instance, Hautala, "Why it was so easy to hack the cameras that took down the web." Chapter 7 also describes the impact of the attack further.

[335] See, for instance, Reuters, "China electronics firm to recall some U.S. products after hacking attack."

actively choose to turn off or leave on particular permissions regarding personal data.[336] Another way to bolster security of one's devices is to conduct an audit. This can be simply done by searching Shodan. This search engine also mentioned in the previous chapter is "the world's first... for Internet-connected devices." It allows users to search for IoT devices. Sometimes criticized as a tool that enables attackers, Shodan in fact enables consumers to verify that their own devices are not publicly accessible beyond necessity. If a search reveals that a private device is widely accessible, then the device's network settings can and should be modified. Shodan hence serves to bolster security when sensibly used.[337]

Still another method by which customers can shore up security of their devices and cloud-accessible services, is to select and maintain strong passwords. Choosing a truly strong password is something of a contentious art, as there are competing perspectives that offer conflicting advice on particulars. Moreover, once valuable guidelines may have long worn out their utility. Nevertheless, some common

---

336 For instance, for both Android operating systems and Apple's iOS, a user can easily change the permissions of each app used. On the iPhone, go to Settings→ Privacy. From here, you can modify which particular apps have access to which particular elements, including for location services, camera access, and access to one's contact information. On Android, go to Settings → Apps → App Permissions for the same functionality.

For certain apps, such as a navigational one, basic functionality may necessitate access to personal data such as location services. For others, such as a memory game, access to the same would strongly imply the use of data mining. Depending on your evaluation of mining's worth to yourself, you can thus choose to turn off such apps' access to certain types of personal data.

337 Shodan itself is linked in the previous chapter's "Further Exploration" section. A strident and representative criticism of the platform can be found in Porup, "'Internet of Things' security is hilariously broken and getting worse." For a quick tutorial on using Shodan, see Miessler, "A Shodan Tutorial and Primer," which was originally written in 2014 but remains relevant. Paul, "How to quickly check that your home IoT devices are secure" provides a methodology for specifically identifying how exposed your IoT devices might be.

recommendations are clear. First of all, not all passwords are equally important. Some accounts may be primary ones if they are linked to more sensitive ones or if they are linked to many other ones. Your primary email account, for instance, would be one such bottleneck of special value. Were a hacker to gain access, he or she could potentially send login credential replacement requests to bank accounts, online shopping accounts, social media accounts, and more; from there, it would be happy hunting to identity and other theft. To forestall some of this impact, prominent service providers, including Google's Gmail and various banks, strongly recommend that users implement "2-step verification." As the name suggests, this security measure requires that users validate a login request through an alternative means associated with the account, such as via text message or phone call. Similarly, universities are beginning to partner with Duo Security to provide similar multi-factor authentication, with hundreds of leading companies joining the thousands already conjoined in April 2018.[338]

For similar reasons, another password best practice is to use several different ones. This way, exposure of one account has a contained impact. Still another set of rules follow from modifications to past practices. In 2003, the National Institute of Standards and Technology (NIST) released version 1.0.1 of "NIST Special Publications 800-63." This document provided guidelines for strong passwords that were not mandated, but came to be widely adopted. Chief among the recommendations was that consumers should implement passwords that consisted of at least one upper case letter, one lower case letter, one number, and one special character, while being at least eight characters long and eschewing common words. It was also recommended that the

[338] For Google's instructions on setting up 2-factor authentication, as well as their justification for doing so, see the "Further Exploration" section of this chapter. Duo's partnership with industry leaders in April 2018 was announced in Duo, "Hundreds of Leading Technology Companies Partner with Duo Security to Provide Zero-Trust Security."

password be changed at least every two years.[339] These parameters would, the authors, believed, result in passwords difficult to deduce.

These sorts of passwords may have been difficult for humans to guess—however, the standard ways in which the recommendations were implemented meant that computers have had a relatively easy time at deducing our secrets, particularly manifesting now with cloud-connected Things. Once again, however, a way to improve the state of play is at hand. NIST release revised password guidelines in 2017 that increase "entropy" to make it more difficult for would-be invaders to deduce passwords. Three key tips from these revisions are to: (1) think of passwords as pass phrases that link ideas, rather than inorganic jumbles of numbers, symbols, and words, (2) choose elements of the phrase unique to your life so that others cannot easily guess it, and (3) leverage the human brain's visualization power to organize and remember your password.[340] Undoubtedly, these guidelines will not be airtight either. Security is based on mathematical and technical bases, but it also fundamentally depends upon human processing. Human proclivities to perhaps only use letters in future passwords, to capitalize the first letter of each new word within the pass phrase, or to rely on particular sorts of word associations, could well mean that within another ten years, NIST will update its guidelines again. However, in the near-term, this new system appears to be a massive improvement, which looks to increase the strength of passwords more than a million-fold.[341]

One further do-it-yourself recommendation of the new NIST guidelines, is use of a password manager. This tool aggregates and organizes all the passwords that a user may have. Through a single

---

[339] See NIST, "Special Publication 800-63 Version 1.0.1," 53.

[340] These three guidelines are summarized at Garcia, "Easy Ways to Build a Better P@$5w0rd." The full suite of updated NIST guidelines is at NIST, "Digital Identity Guidelines: Now Available."

[341] As illustrated in Garcia, "Easy Ways."

password to access this manager, a user can then easily access all of his passwords. Various security experts have championed the use of password managers in the past, citing their ease of use and volume storage; there are, moreover, numerous cloud-based managers.[342] Regardless of where they store passwords, these managers do, however, raise some security concerns. Particularly, they concentrate vast power in a single location. This creates a single target which contains access to essentially all of a user's most confidential information. This ripe target may be pursued by criminals or cyberterrorists; it may alternatively be pursued by greedy employees of the company itself. Regardless of acceptance of the password manager, however, it is clear that a variety of viable and simple methods for increased security are available to consumers.

Other attacks have been described earlier and in Chapter 7, but they nonetheless do provide valuable lessons that further support how addressable present vulnerabilities are. For instance, in the WannaCry attack from May 2017, scores of hospitals were afflicted by ransomware. Individuals' lives were literally placed in jeopardy. A terrifying incident, it netted more than $100,000 for the perpetrators, money paid through Bitcoin.[343] Yet, the incident was enabled by the failure of administrators to either switch to a more current operating system, or implement software patches that had been released months prior to the attack, and

[342] Bruce Schneier, in his prominent blog "Schneier on Security" and elsewhere, has been a strong advocate of password managers. See Schneier, "Security of Password Managers" for one discussion of their virtues. Some prominent password managers as of early 2018 are identified in Gewirtz, "The best password managers for 2018." As described there cloud-based managers include Blur, StickyPassword, and LastPass.

[343] Collins, "Victims of the WannaCry Ransomware Attacks Have Stopped Paying Up."

recommended for national adoption.[344] If the recommendations had simply been followed, the effects would have been mitigated—likewise for the massive Equifax breach, the impact of which was similarly enhanced because security patches had not been applied.[345]

An earlier attack indicates another potential source of increased security from a technical standpoint. In 2009, data broker RockYou was the target of another large-scale attack. RockYou partners with smartphone game developers to provide marketing and other information that might be of value. As such, it maintains a massive database of customer information. A hacker was able to enter this repository by exploiting a common—and correctable—flaw in the company's SQL design code. Once inside, the intruder was able to access the unencrypted information which included usernames and associated passwords in plaintext, in turn enabling access to email credentials and more. All told, at least thirty-two million accounts were compromised.[346] A core lesson from this attack is that encryption should likely be deployed more broadly.

Encryption is the process of encoding information so that only a particular key can effectively decode it. While it can serve to protect information even in the event that a database is breached, it does come with its own drawbacks. For instance, it makes information access more complicated than it would be otherwise, potentially diluting some of the efficiency benefits that the cloud promises for information sharing. Moreover, encrypted files do not in themselves help to address

---

[344] WannaCry was a global attack, but primarily affected hospitals in the United Kingdom. In response, the UK government's National Audit Office released a comprehensive report in October 2017 detailing the facts of the attack and lessons learned. See the "Further Exploration" section of this chapter for this report.

[345] See Knight, "Equifax reveals full extent of last year's massive data breach," and this chapter's "Further Exploration" for more on the Equifax breach and its preventability.

[346] See Cubrilovic, "RockYou Hack: From Bad to Worse."

ransomware attacks, since even if the attackers are unable to decrypt the information within, they can still prevent legitimate users from accessing the files.[347] Encryption will hence not be suitable in every situation, but undoubtedly, it could be implemented more widely to increase security. As Gemalto's Breach Level Index has identified, of the more than nine billion breaches in the past six years, only four percent were "secure breaches," where encrypted data was stolen.[348] Increasing use of encryption would certainly help to reduce this number.

All of the techniques highlighted here are relatively simple to implement. Moreover, all could have mitigated, if not prevented, massive data breaches of the recent past. Despite belabored cries that the sky is falling, there has undoubtedly been improvement in how security is valued. In 2012, the Ponemon Institute released a report based on 584 then-recent data breaches. Its researchers found that in an enormous 34% of the cases, the root cause involved a negligent insider. Particularly, as well, an additional 6% of attacks resulted from employees' failure to shred confidential documents.[349] The percentage of attacks involving internal actors, as noted above, dwindled to 23%. This percentage drop is not simply the result of an increased number of external attacks, but also attests to a growing focus, however slowly, upon security. Also promising from these studies over time is a reduction in ignorance: in the 2012 study, the Ponemon Institute found that in 44% of the breaches the root cause or motivation remained uncertain; by contrast, only 11%

[347] A quick visual summary of encryption is available at ICO, "Types of encryption" along with a report of further lessons to be learned from recent cyberattacks, as compiled by the United Kingdom's data protection authority. Some more positive and negative aspects of encryption are addressed in Stieglitz, "Encryption: Pros and Cons." Encryption is part of a larger discipline known as cryptography. For a rich history of this discipline unto the twentieth century, see David Kahn's 1967 book *The Codebreakers*, updated in 1996.

[348] Gemalto, "Data Breach Statistics – Breach Level Index."

[349] See Ponemon Institute, "Aftermath of a Data Breach Study."

of recent attacks in 2017 were found to be so.[350] This is significant, since only by eliminating ignorance can companies gain a better handle on present vulnerabilities and identify viable solutions.

SPOTLIGHT: WHO ARE MULTINATIONAL CORPORATIONS COMPELLED TO SERVE?

The United States is not the only nation acting to establish clear rules for e-discovery of data stored on the cloud and stored abroad. In fact, the European Union currently is considering a Regulation (immediately binding on member countries) and Directive (requires member countries to each implement their own version of the legislation) that would largely mirror the CLOUD Act. The EU E-Evidence Proposal would allow a judicial authority in one nation (at the urging of law enforcement) to oblige service providers based out of another to provide electronic evidence such as emails or text messages within ten days, or six hours in emergency circumstances. This compares quite aggressively with the ten months presently allotted via Mutual Legal Assistance Treaties (MLAT), a timeframe that has frustrated law enforcement worldwide and motivated passage of the CLOUD Act.

Via the proposal, service providers also have some recourse to challenge information requests that they believe are unjust. What do you think of increased global clout by law enforcement? While the Proposal has already been through a period of public review, it still needs to pass Parliament, a process which will likely take at least a year, if not longer.

More information is at European Commission, "E-evidence."

As with development of the IoT, we can expect that the best cloud services will be the ones that satisfy customer needs, and security is a significant aspect of this. Naturally, we can expect that the market

[350] DesJardins, "Why Hackers Hack: Motives Behind Cyberattacks."

should shift to produce more security-conscious businesses. Perhaps the rate of security progress remains insufficient, but there remains reason to be optimistic. Clearly both companies and consumers stand to benefit by reducing the risk of cyberattack. For individuals of course, their personal data in addition to the devices they rely upon become strengthened. For companies, cyberattacks can massively damage reputation in addition to risking costs for consumers' damages, and from regulators. Both consumers and corporations hence need strong security, and legitimate cooperation seems to be what is in order rather than stunting progress to the detriment of all parties. Fundamentally, this cooperation could be realized through improved education efforts and simpler means by which consumers can manage security.

Thus, overhaul is not necessary. More attention certainly is required, but this is something that can be addressed, such as through increased corporate efforts at education, and transparency. With its ubiquitous, standardized nature, the cloud hence offers a prime method for customer education—a plausible solution. The costs of cybersecurity attacks are massive for the corporations involved; they may have to pay damages, they may have to replace compromised infrastructure, they may moreover lose customer support.[351] It is in customer's and corporations' best interests to bolster security.

Additionally, as a tool, the cloud offers another specific way of bolstering security: startups and others can outsource security to experienced professionals via the cloud. As such, these outsourcers can

---

[351] At least one case indicates that attuned customers will punish companies with which they are displeased regarding privacy or security concerns. In August 2018, media interface Plex announced a new privacy policy to begin in September that would require collection of user data and permit its sale to third parties. Customers responded vocally on Reddit and Plex's own forums, threatening to end their patronage. For its part, Plex responded by promising changes as soon as possible, and reiterating that the company had "ZERO interest" in learning what content was contained within users' libraries. For some more on this incident, see Perez, "Plex changes its new privacy policy after backlash, clarified it's not trying to see what's in your library."

avoid trying to implement a system from scratch, a process that would likely result in unintended gaps. As with many other areas of cloud development, growth of security services looks rosy. Gartner estimates, for instance, that cloud security is progressing at a faster rate than the information security market in general, and will approach nine billion dollars by 2020, up from its $5.8 billion in 2017. This growth is anticipated for all aspects of cloud security: from implementation of firewalls to user authentication to remote vulnerability assessments and more. As such, products are diversifying, and well-respected players including Amazon, Symantec, and McAfee are continuing development.[352] An offshoot benefit of cloud use as a whole further benefits security: the cost-saving made possible by other aspects of cloud services mean that more money can be invested in security measures.[353] Security can hence become a greater focus moving forward.[354]

As many of the most serious cyberattacks indicate, the major vulnerabilities today can be addressed. A wide array of do-it-yourself techniques allow customers to themselves bolster security of their products on and using the cloud. Moreover, backend and infrastructural changes are being implemented in corporate products, indicating that the security threat is being taken quite seriously. It is is in all legitimate

352 See Gartner, "Gartner Forecasts Worldwide Cloud-Based Security Services to Grow 21 Percent in 2017" for projected numbers on cloud security growth. For more on key players see, for instance, Aspa, "Top Cloud Security Companies."

353 Cyberattacks can be extremely expensive, and the financial cost of damages can understandably vary wildly. One estimate by the Ponemon Institute in 2016 found that data breaches cost $4 million on average: Ponemon Institute, "2016 Cost of Data Breach Study: Global Analysis." In its 2018 study, linked in this chapter's "Further Exploration" section, the Ponemon Institute found that the average cost of a data breach was $4.43 million, while a substantially reduced $2.88 million for companies that deployed security automation, such as could be provided by cloud services.

354 For a short compilation of still more cloud security benefits, see Ruff, "7 Reasons why cloud-based security makes sense."

parties' interests to improve security and move quickly away from a state of avoidable vulnerabilities. This path forward will require increased collaboration between IoT and cloud service providers as well as consumers, and can take a number of forms. For instance, perhaps products should come with a required online course to teach about basic security mechanisms, and perhaps manufacturers should focus on design more intuitive interfaces by which consumers can adjust security settings. Regardless of the specific approach taken, the cause for many cyberattacks is not the cloud itself. We should propel the technology forward to realize its sunny benefits all the sooner.

The Present Environment is Opportune

Besides the promise the cloud holds for bolstering cybersecurity, the technology is moreover developing within a legitimate and promising environment. Like the Internet of Things, like drones, like smartphones in the early noughties, the cloud is still in its infancy. How things appear now is not necessarily how things will be in five years or even one.

Yet, for what there is now, there remains choice. Presumably, the cloud's growth is not the result of Stockholm Syndrome, captive consumers feeding an indomitable beast. Rather, the scale of growth indicates genuine excitement. Startups, government, and individual consumers are all already streamlining their activities, increasing efficiency, and gaining new insights. The choice is reflected as well in the massive variety of services offered. Consider, for instance, AWS which offers more than a hundred different services through the cloud, many of which are readily deployed by startups. It and competitors also offer various tiers of business models, with high scalability.

It seems sensible for individuals to have control of sensitive data. This control, however, should also manifest in genuine autonomy: individuals should also be trusted to deal their sensitive data as they see fit. Privacy like all rights is grounded in a fundamental respect for individuals; this respect entails freedom of choice. Moreover, when it comes to behavioral and other sort of data that may be collected through cloud services, the dangers are often overstated. As discussed in the latter half of Chapter 3, the informativeness of this data appears to be overstated. Additionally, the tradeoffs are valuable. Discussed at greater length in Chapter 4, the free or nearly-free services provided by leading tech companies, including cloud providers, are subsidized through collection of such behavioral data. It would be unfair and not economical for these services to remain without receiving something of value from users. Allowing consumers to completely control their personal data would hence not be tenable within the old, familiar ecosystem. Instead, some changes would need to be made, such as increasing service costs and perhaps restricting content for those without a subscription. This is neither good nor bad, but it is a change that customers should be aware of, and ideally, should be a decision that each individual can make.[355] Unilaterally forcing everyone into a system of payments would be a step back for autonomy and respect.

Data ownership and use of what is processed through the cloud, remain things tenuous and imprecisely defined. Nevertheless, we should trust that the laws will adapt appropriately. This has been the case in the past. For instance, telegraphs and telephones provided new methods by which law enforcement could surveil citizens, and wiretap protections developed in time, with 1967's *Katz v. United States* and the Federal Wiretap Statute one year later. Cable and satellite developed over time, protected in part by the 1986 Electronic Communications Privacy Act. The cloud is certainly a revolutionary technology which is rebuilding national infrastructure, communications, and the economy. As such, it will undoubtedly have massive impact, perhaps more than prior technologies. While this power to influence may raise some risks which critics have seized upon, the power also means that the benefits will be

---

355 Some, such as Downes, "GDPR and the End of the Internet's Bargain," and Tiku, "Europe's New Privacy Law Will Change the Web, and More," have speculated that the EU GDPR will usher in ecosystem changes such as are described here. This remains to be seen as the law is implemented in the second half of 2018 and beyond.

magnified. In realizing these benefits to efficiency, equality, and more, we should take the example of flexible law in the United States as our guide. America exists through a Declaration of Independence almost two hundred and fifty years old. Many societal aspects have changed since then, with increased opportunities and improved technologies. Yet, the core laws remain relevant because they left leeway for interpretation, avoiding limitations unless they seemed absolutely necessary.[356] Promising a new sort of revolution, the cloud should not be constrained, especially when the risks identified are largely speculative.

In addition to this natural settling of the law that we can expect, the progression of events recently indicates a growing focus on privacy. Consider the desolate desert of cyberspace legislation just some twenty years ago. Since then, the FTC began punishing companies which changed their privacy policies with little or no notice to customers.[357] Struck by severe costs and public relations damage, companies have also made a concerted push towards transparency, led by large tech companies including cloud providers Google, Amazon, and IBM. Now, some laws, particularly the CLOUD Act and the EU GDPR are beginning to consolidate the nuanced rules concerning data ownership. This vision of consolidation is clearly upon the minds of industry leaders.

---

[356] Law's flexibility has been described as important to all interpreters, regardless of conservative or liberal ideology. For a longer discussion of this idea, see *The Economist*, "Why the American legal system is so flexible," itself a review of Stephen Presser's book *Law Professors: Three Centuries of Shaping American Law*.

[357] In 2002, after a public data breach, the FTC required Eli Lilly to develop and maintain an appropriate security program for customer data: FTC, "Eli Lilly Settles FTC Charges." In 2004, the FTC fined Gateway Learning Corporation due to retroactive changes to its privacy policy, and required changes to the policy: FTC, "Gateway Learning Settles FTC Privacy Charges." In 2011, Facebook settled with the agency to avoid charges that it had improperly protected private customer data: FTC, "Facebook Settles FTC Charges." This last settlement has been cast in doubt after the Cambridge Analytica scandal widely broke in 2018, and, as addressed in Shields and Quinn, "Facebook Could Be Fined," the social network may soon find slapped with further penalties.

As Apple, Facebook, Google, Microsoft, and Oath noted in their letter of support for the CLOUD Act, "Our companies have long advocated for international agreements and global solutions to protect our customers and Internet users around the world."[358] Confusion benefits no legitimate party involved in the cloud or other aspects of data processing. Nevertheless, patience and trust in the ecosystem will best lead us to the benefits ahead without shutting off possibilities.

Undoubtedly, a balance needs to be struck between individuals' autonomy and economic success. As trends and history suggest, we should allow the ecosystem to grow of its own accord, rather than mold it to regulators' necessarily limited vision. For the known benefits at hand, as well as those as-yet unknown, we should field the cloud sensibly along its present path. Nurtured by the sweat of practitioners, the cloud will expand into its revolutionary and massive potential. In short time, it will lead to a smarter, more efficient, and more enlightened society.

In Conclusion

A still-emerging technology, cloud computing boasts society-altering capabilities. These benefits and risks include issues on the cybersecurity front as well as assumptions about the ownership of personal data. International laws and regulations acting upon data in the cloud also remain in a nascent state, with competing rulings and vagaries of international law inhibiting both expansive and constrictive developments. The present hazy period, with case resolutions pending presents an opportunity to seriously evaluate the direction in which we want to push cloud computing. These evaluations on personal, societal, and global scales will affect decisions both minor—such as Ava's purchase of data storage—and massive—such as the security and infrastructure of nations.

---

358 See Hatmaker, "As the CLOUD Act sneaks into the omnibus" for a transcript of the full letter.

Further Exploration

The Privacy Rights Clearinghouse compiles a list of publicly disclosed breaches and the number of affected records since 2005: **https://www.privacyrights.org/data-breaches**. Tracking since 2013, the **Breach Level Index** offers another account of breached records, along with a "Risk Calculator," summary reports, and other tools: **http://breachlevelindex.com**.

Founded in 2002, the **Ponemon Institute** is a highly reputed research center focused on data protection and information security. The company's **2018 "Cost of a Data Breach"** report addresses costs of incident discovery and response across the globe. It can be accessed via **https://databreachcalculator.mybluemix.net**.

Developments in the **2017 Equifax data breach** have been charted by the FTC at **https://www.ftc.gov/equifax-data-breach**, particularly focused on some recommendations for action. Brian Krebs, a well-respected journalist in cybersecurity, provides an in-depth analysis of the event on his blog **"Krebs on Security."** This is accessible at **https://krebsonsecurity.com/2017/09/equifax-breach-setting-the-record-straight**. A May 2018 retrospective by Equifax, obligated by the United States Securities and Exchange Commission, provides a more recent update:
https://www.sec.gov/Archives/edgar/data/33185/000119312518154706/0001193125-18-154706-index.htm.

The annual **Verizon Data Breach Investigations Report (DBIR)** offers a thorough compendium of annual cyberattacks, some of which have not been publicly disclosed. It is available through registration as a "Verizon Insider" at **http://www.verizonenterprise.com/verizon-insights-lab/dbir**.

The **Massachusetts state website** provides additional information on common cyberattacks. Linked within are tips for attack prevention: **https://www.mass.gov/service-details/know-the-types-of-cyber-threats**. The **US-based National Cyber Security Ventures Corporation** site also provides education, investment, reports, and numerous other services to defend against cybercrime: **www.nationalcybersecurity.com**.

Multi-factor authentication is a security measure gaining traction. **Google** provides instructions on how users can install it at **https://www.google.com/landing/2step**. The world's leader for multifactor authentication, **Duo** provides the service for thousands of companies across the globe. Further information can be found at **https://duo.com/product/trusted-users/two-factor-authentication**.

Password managers are another potential means of increased security, though come with their own risks, such as in creating a vulnerable target. A compilation of managers, recommended as of early 2018, is at **https://www.cnet.com/news/the-best-password-managers-directory**.

Following the 2017 **WannaCry** ransomware attack, **the UK's National Audit Office** (NAO) released a summary report of the incident, response, and lessons learned. It is available at **https://www.nao.org.uk/report/investigation-wannacry-cyber-attack-and-the-nhs**.

Official background **on European Union General Data Protection Directive** (EU GDPR) is available at **https://ec.europa.eu/info/law/law-topic/data-protection/reform/rules-business-and-organisations/principles-gdpr_en**. The site also features toolkits and other resources to aid with compliance. An alternative crowdsourced portal of information is at **https://www.eugdpr.org**, and a clean summary of the Regulation's 99 Articles plus 173 explanatory Recitals has been compiled by **intersoft consulting services: https://gdpr-info.eu**.

In March 2018, the United States passed **the CLOUD Act** which, enables easier access of consumers' communication data by law enforcement, even if stored abroad. The full text can be found under **Division V of the Consolidated Appropriations Act of 2018**, accessible at **https://www.congress.gov/bill/115th-congress/house-bill/1625/text**.

The United States has entered agreements with Switzerland and the European Union, which are known as **Privacy Shields**. These govern transatlantic commerce with more information available at **https://www.privacyshield.gov**.

The **NOAA Big Data Project** is a collaboration between the United States' National Oceanic and Atmospheric Administration and cloud providers IBM, AWS, and Google Cloud. The tech companies agreed to build the cloud infrastructure to host NOAA's big data at no cost, and through this project, NOAA seeks to improve "the accessibility of data while also maintaining a fair and level playing field for everyone." More information on the project, as well as links to the collected data, is available at **http://www.noaa.gov/big-data-project**.

Amazon Web Services (AWS) offers SaaS as well as other cloud applications: **https://aws.amazon.com**. **Google Cloud**, for developers and others, is at **https://cloud.google.com**. **Microsoft's** popular cloud computing offering, **Azure**, along with the company's sales pitch, can be found at **https://azure.microsoft.com/en-us**.

Chapter 9: Genomic Privacy

Key Questions:

- What does it mean to be human?
- How can genomic sequencing be done today and who can do it?
- What do personal genomics companies promise? Do those promises hold up to what they deliver?
- Who has access to sequenced genetic data? Who should have access?
- How secure are the results of personal genomics services?
- What does biometric data actually reveal? What does it not reveal?
- Who, if anyone, owns genetic data?
- What legislation presently exists to preserve genomic privacy? What are its limitations, and what gaps, if any, need to be filled?

Consider: Two weeks ago, Juliana signed up for a genetic profiling service. Offered by the startup Gateway to Tomorrow, it has recruited several thousand customers across the United States, encouraging the public to "Unwind. Live Better."

As described on promotional information, the program sequences key segments of a customer's genome to identify an individual's likelihood for various diseases. The program also quantifies the probability of biologically neutral traits, such as whether a participant carries dominant genes for large or small ears. Moreover, to generate sales, the program further advertises that it seeks an ever-growing number of participants (for a relatively modest $500 fee), so as to fine tune its predictive capabilities and unlock even more information for participant genomes in the future.

In turn, the company promises to subsidize additional tests for every five new people that a customer recommends. The company also acknowledges that the kits are not actual medical devices, and furthermore that any diagnoses are probabilistic rather than determinate. Juliana, under some pressure from her cousins who also signed up for the service, finally decided to participate, mailing in a vial of her saliva for sequencing.

Earlier this week, Juliana logged in to discover her results: apparently, she has a "sixty percent" risk for brain cancer. Perhaps the findings are in error. Perhaps not. Her great-grandmother also suffered from brain cancer, but Juliana had not thought herself at risk as well, nor had any previous physicians warned her of the risk. The more she thinks about it, however, Juliana can believe that such a risk is possible, even likely. In turn, Juliana is contemplating radical changes to her lifestyle. Among other modifications, she is contemplating severely limiting her use of smart devices, since a few prominent studies suggest radiation from electronic devices is a minor cancer risk.

Should Juliana have had access to the personalized sequencing technology? What should her next steps be?

—Inspired by personal genomics diagnoses of the recent past

Since the dawn of human existence, we have wondered what it is that makes us human. Is it a capacity to reason: to learn from our mistakes and invent gadgets and ideas? Is it an ability to empathize: to comfort, weep, and grieve? Is it possession of some non-corporeal soul? Is it a particular mechanical configuration of body parts: two arms, two legs, ten toes, ten fingers, two ears, two eyes, a mouth, and a nose?

No answer has yet been wholly satisfactory. Each excludes, it seems, some of those who ought to be considered kin. For instance, children and those mentally retarded often display stunted reasoning, while every person is at least at times guilty of inconsistent and irrational behavior. Emotional intelligence also varies wildly, with sociopaths and psychopaths who utterly lack empathetic feeling. Moreover, soldiers lose limbs while children can be born with extra digits.

The question of identity bears more than pedantic significance. In a highly populated world of limited resources, bound by limited time, certain creatures inevitably receive preferential treatment. In the modern political space, moral "personhood" serves as the basis for basic rights. Facetious definitions of "person" have had devastating consequences in the past: the view of dark-skinned Africans as primitive and subhuman, for instance, allowed their enslavement around the world. In Nazi Germany, Hitler and his propaganda ministry broadcast the superiority of the Aryan race and imperfection of all others, especially Jews, homosexuals, and gypsies. Shortly before the Rwandan genocide in 1994, the Hutus described the Tutsis, a rival ethnic group, as cockroaches, stripping them of their humanity. In the future, as artificial intelligence develops robust capacities that rival humans, we may again be forced to expand our conception of what receives privileged attention in order to develop a more equitable society.

Faced with the baffling question of "What is a person?" some have attempted to shuffle our ordinary conceptions of what fits together. This has been the tactic adopted by several prominent moral philosophers. John Stuart Mill and Immanuel Kant, for instance, are often considered to be on extremes of the moral continuum, with Mill's high valuation of an action's consequences, compared to Kant's adherence to a rule for action, regardless of consequences. Nevertheless, the thinkers both suggested that being a "person" requires possession of some minimal mental capacity. For Mill, those who cannot feel pain, and for Kant, those incapable of rational thought, are not persons and may not possess the same rights as the rest of us. Yet, the emotions and care many of us feel towards those mentally and physically divergent from ourselves come in cold conflict with such convenient classifications. Scholars, artists, and researchers continue to push on to uncover the shared soul of our humanity.

Alternatively, another attempt at classification has dug deeper into the physical body. In biology, a species includes organisms that can reproduce and create fertile offspring. Of course, even this classification for *Homo sapiens* is imperfect, given that some become and others are born infertile. Nevertheless, the physical push has resulted in exciting discoveries with massive implications for healthcare.

It began in the 1830s, when naturalist Charles Darwin voyaged upon the *Beagle* around the world. Documenting differences between apparently similar species from different environments, Darwin formulated a controversial theory of evolution. Creatures, he suggested, adapted to their surroundings and over time, those creatures with favorable traits for survival and hence reproduction, would dominate. If external conditions were divergent for two populations of the same species and those species were moreover isolated—as was the case in the diverse Galápagos Islands—then those populations could in turn develop into new species. This was the case with the Galápagos finches.

Monk Gregor Mendel's work with pea plants, though largely conducted in isolation and ignored until the turn of the century, added more to the theory. His controlled studies of pea plant breeding along with meticulous documentation provided evidence that at least certain traits propagated in repeatable patterns. These traits, he speculated, were passed on through discrete inherited units within the organism, units that would come to be termed "genes." Equipped with more powerful technologies such as electron microscopes, X-rays, and hermetically secured lab spaces, subsequent researchers dug deeper to identify the actual mechanisms of gene inheritance.

In the early 1900s, Albrecht Kossel discovered four main nucleotide bases involved in the internal code for traits: adenosine, thymine, cytosine, and guanine. Corroborative work by a group of scientists including Alfred Hershey, Martha Chase, Oswald Avery, Colin MacLeod, and Maclyn McCarty through the early 1950s, indicated that chemically, this internal code consisted of nucleic acids, rather than proteins, carbohydrates, or lipids. Based on the observed structure of these elements, the genetic code came to be termed deoxyribonucleic acid (DNA). Incorporating the work of his contemporaries, by 1950, Erwin Chargaff identified that the four fundamental nucleotide base pairs would only combine in a particular way—adenosine with thymine, and cytosine with guanine—vastly simplifying relevant possible models.

Those models still remained confounding until 1953. At last, a group of British and American researchers including Francis Crick, Rosalind Franklin, Linus Pauling, and James Watson identified that the genetic code wore a double helix structure. This realization rapidly unlocked many more. It became clearer how DNA "unzips" during the process of cell division to replicate and pass on encoded genes. Moreover, by the 1960s, Marshall Nirenberg, Har Gobind Khorana and others built on the prior foundation to fill in the gaps between DNA and traits: they identified that DNA nucleotides are "read" in groups of three, by proteins in the body. Each of these triads codes for an element of a new protein, which in turn provide the means by which the body can perform its essential functions.

Other technological developments have tied these theoretical discoveries more directly to patient care. One essential discovery has found that human DNA typically consists of twenty-three paired chromosomes, including one pair of "sex chromosomes" that determine whether an individual is biologically male or female. Abnormalities in chromosome number that may occur during cell division can tend to result in physical or developmental abnormalities. For instance, trisomy 21, a triplicate of the twenty-first chromosome, results in Down Syndrome. Other developments have made the prospects of genetic sequencing increasingly efficient. At the turn of the twenty-first century, this led a large group of researchers from both the private and public sectors to pursue the first complete sequencing of a human. The Human Genome Project was completed in 2003 and celebrated as a new landmark. Yet, while the project has provided large sums of data that continue to be explored, it has also highlighted the incompleteness of our current genetic understanding. In itself, the genomic data has been unable to precisely determine the likelihood of developing cancer and most other prominent diseases.[359]

Today, researchers recognize that the genome requires auxiliary data to be indicative. Copious non-genomic data, including family history and exercise habits, help to construct a more accurate probabilistic picture. In turn, research has continued to explore other factors besides the genome: epigenetics (literally, "above genetics") studies how environmental and dietary factors can cause isolated physical changes to our DNA, such as the addition of methyl groups to cytosine molecules, modification of the charge of the histone proteins around which DNA is wound, and exposure to infectious proteins known as prions that can attach to base pairs and suppress gene expression.[360] Moreover, research organizations affiliated with the Human Genome Project, such as the Broad Institute, continue to pore

---

[359] There have been many discoveries in the field of genetics, and research initially thought inconspicuous may well turn out to be fundamental and revolutionary, as was the fate with Gregor Mendel's inheritance. Links to further timelines of influential events in genomics are provided in this chapter's "Further Exploration" section—one expanding on the events described here, and another more technical in nature.

[360] See, for instance, *Nature,* "Beyond the Genome" for more on the development of our understanding of epigenetics.

through the massive amount of data previously collected, seeking to uncover decisive connections.

Since the Human Genome Project, sequencing has continued to become a cheaper and far swifter process. Compared to the thirteen-year pioneering endeavor that cost upwards of one billion dollars, full sequencing can now be performed within two days and for less than five thousand dollars.[361] Correspondingly, many companies have sought to capitalize on this increased efficiency and provide personalized medicine as a service to consumers. For a few hundred dollars, companies such as 23andMe, Pathway Genomics, and Ancestry.com offer to sequence any participant's genome and unlock predictions that individuals would otherwise never access. Post-sequencing, results are available to the individual, and are also compiled in aggregate for further analysis by the companies. The service they offer is called *personal genomics*.

Recent regulation, to be discussed below, prohibits explicit diagnoses of serious diseases via personal genomics on account of consumer protection. Nevertheless, diagnoses of other traits are permitted and the service is now available to anyone with means. In addition to for-profit companies like 23&Me or the fictional Gateway to Tomorrow, others such as deCODE Genomics have added variations to the model; having chosen a set number of participants at the programs' start, they use repeated check-ins to analyze epigenetic and other changes over time.[362] These companies, whether private or public, stand to improve our current understanding of the genome and radically improve medicine in the future. The private companies can also make sizable profits in the present.

As with the other technologies discussed in this book, personal genomics is not pure sunshine. In a variety of ways, personal genomics potentially threatens individuals' privacy in the short and long-term.

[361] Lewis, T, "Human Genome Project Marks 10th Anniversary."

[362] For some links to genetic profiling companies, see the "Further Exploration" portion of this chapter.

Databases that store genomes are bound to be tempting targets and, even more so than other data repositories, ones that cannot be decisively secured. Moreover, future legalized genetic discrimination lurks as a storm cloud on the horizon, and data collected now may help to enable prejudicial policies. Immediately, as well, misdiagnoses and potential misinterpretation by inexperienced customers can result in customer overreactions—of the sort Juliana is considering—or else false confidence enabling risky behavior.

On the other hand, though our current understanding of causal relationships involving genetic data may be in a nascent state, the conclusions still promise fundamental insight to cure disease and improve customers' quality of life. Depriving present and future patients of access to beneficial knowledge would arguably be immoral. Moreover, research restrictions might also push the research underground, in turn amplifying abuse and muzzling benefit.

Whatever part of our identities that genomes may be, they are at the core of cutting-edge research and products. Personal genomics is one of these products, recently emerged for wide consumption. Should it be allowed? Should it be regulated further? Undoubtedly, how it develops will have enormous impact for our health and society.

A Case For Increased Regulation of Personal Genomics

Due to increased regulation by the FDA, Juliana's situation is less likely to occur in the United States today than in the recent past. At least through the end of 2017, personal genomics companies are prohibited from diagnosing many severe diseases, including cancer. This prohibition is a recognition that present support structures are inadequate, and also a recognition that the risk of misdiagnosis is too high. Nonetheless, recent moves threaten to eliminate these essential protections, while other dangers for promoting personal genomics have yet to be addressed.

The risks from the availability of personal genomics are manifold. First, though the full extent of what is being revealed is as yet unknown, it is undoubtedly immense and reveals sensitive information. Hence, genomic data that is broadcast and stored, as it is in the existing models, threatens to haunt future generations, while remaining a cybersecurity threat in the present. The likelihood of future exploitation is enhanced by the profit-maximizing mentality of the personal genomics industry and its key associates: pharmaceutical and insurance companies. Focus on financial profits comes at the expense of citizen autonomy, and hence indicating that the present custodians of genomic data are not trustworthy. Moreover, the incomplete nature of genomic knowledge is highly dangerous. While there undoubtedly is genuine insight buried within, personal genomics presently may be fueled by dubious results, a product of profit-maximization schemes. Even with recent regulations, the broad freedom afforded to personal genomics companies allows them to promise more than they truly can provide, and hence push participants towards harmful action.

For all of these reasons, the current climate is not one conducive to ethical and truly valuable use of personal genomics for consumers. The substantial risks of personal genomics for privacy and other civil rights must be confronted now, and nipped in the bud. Otherwise, corrupt influence will continue to spread and stunt the market's potential. Serious intervention needs to occur, and until it does, personal genomics should not be permitted to continue.

Leakage of the Human Genome

For each of us, our genome is a predetermined sequence that codes our physical and mental development. It dictates how our bodies initially develop and continues to inform how our cells reproduce and replenish. The genome instructs smaller molecules called proteins on how to behave—when they are to be born, when and how to activate, and when to dissipate. The resulting processes induce development of the body, increasing longevity, as in immune defense against disease, or, as in the development of a cancer, ushering in the end.

The genome is not the sole determinant of development, as identical twins raised in different environments clearly reveal.[363] Nevertheless, the genome does determine a great deal, and includes markers that have been shown to encode valuable information. For instance, the genome determines biological sex. Humans typically have twenty-three pairs of chromosomes. The twenty-third dictates gender: members of both biological genders have an "X" chromosome, but whereas females have a second X chromosome, males have a "Y" of substantially shorter length. Moreover, chromosomal aberrations, such as the presence of an extra chromosome, indicate whether an individual has a developmental disability: trisomy 21 (a triplicate of chromosome 21), for instance, results in Down syndrome.[364]

From the particular sequence of our genomes, other traits are also determined. Eye color, hair color, ear lobe type, and more have been linked to particular chromosome segments; in fact, services like 23&Me strive to identify these phenotypes and in turn quantify their likelihood of hereditability.[365] At face value, these sorts of traits may not

---

[363] For instance, McKie, "Why do identical twins end up having such different lives?" points to some cases of divergence. Additionally, a large-scale study called E-Risk was conducted on 1,116 pairs of same-sex twins born in the UK in the mid-90s. The full dataset is not public, but the study designers have published a few papers on the material, with more likely forthcoming: Bates, "Identical Twins, Different Lives."

[364] Mutations in the sex chromosomes in particular can result in large internal and external abnormalities. The World Health Organization dives into some of these intricacies at World Health Organization, "Gender and Genetics."

[365] For instance, blond and red hair have both been decisively linked to very specific segments of the genome: Pennisi, "The Genetics of Blond Hair" and Lallanilla, "Are You Carrying the Red Hair Gene?" Eye color has also been linked to decisive segments of the genome: *Cell Press*, "Show Me Your DNA." From studies of mice and men, ear lobe type has also been clearly associated with the genome: Adhikari, et al., "A genome-wide association study."

seem particularly private—they are, after all, overtly visible. Nevertheless, they are identifying characteristics and, especially with the growth of biometrics for security, their revelation is still needless exposure.

Other encoded data is undeniably more sensitive. Degenerative conditions, such as Huntington's disease have been diagnosed from DNA. Similarly, studies conducted by an international group of researchers and published in 2017 suggest that Alzheimer's and Parkinson's can be predicted by analyzing a combination of genetic markers spread through the genome.[366] Knowledge of proclivity to such diseases can help affected individuals and their loved ones in treatment. However, if publicly released, such information could also be used against the individuals whose data is encoded.

The Genetic Information Nondiscrimination Act (GINA) offers some protections against employment and health insurance discrimination. It was passed in 2008 by the United States Congress "[t]o prohibit discrimination on the basis of genetic information with respect to health insurance and employment."[367] The bill cited a past history of local discrimination. This history includes eugenics, the practice of actively selecting for particular traits, perhaps most prominently used by the Nazi Empire during World War II. Through genetic discrimination, governments and businesses have used the presence of "undesirable" traits as the basis for limiting the concept of "person"—relegating some individuals to a lower tier, if not excluding them entirely.

Among other restrictions, GINA prohibits employers from purchasing employee information from medical databases. However,

---

[366]See, for instance, National Human Genome Research Institute, "Learning About Huntington's Disease;" Devlin, "New Alzheimer's test can predict age;" and Chang, "A meta-analysis of genome-wide association."

[367] See the "Further Exploration" section of this chapter for a link to GINA's full text and official supporting materials.

several open projects, noble though their purpose may seem, still threaten the carelessly generous. One of these, the 1000 Genomes Project, aimed to produce the largest public catalogue of human variation and genotype data up to that point. It ran from 2008 to 2015 and through the public release of data, its leaders hoped to usher in new discoveries. Though closed to the public-at-large, the results were made available to authorized scientists—or, more accurately, users with the credentials of authorized scientists. One unintended consequence of the database was re-discovery of participant identities. In a 2013 article, Dr. Gymrek and a group of researchers similar to Dr. Sweeney's prior efforts, described how they were able to identify the surnames of males involved with high precision by analyzing portions of the sequenced Y chromosome and information from public genealogical databases. This was achieved despite the program supposedly protecting the anonymity of its participants.[368]

Projects like the 1000 Genomes Project nobly aspire to benefit human understanding and improve health services. Yet, this hope is far too hazy, especially when compared with the possibility of genetic discrimination. In March 2017, the US Congress considered a bill (H.R.1313) to allow employers to use health data as a basis for determining which employees received benefits.[369] As articulated, employees would be required to submit a genetic test or else pay a higher rate for health insurance. Under GINA, discrimination by employers on the basis of genetic information is prohibited save for six exceptional cases. One of these exceptions is when an employer offers a wellness program, an employee provides voluntary consent, disclosure of personally identifying data is limited, and any data disclosed to the

---

[368] Gymrek et al., "Identifying Personal Genomes by Surname Inference." The data from the project and more on the project scope is available at http://www.internationalgenome.org/about.

[369] H.R. 1313 can be found at https://www.congress.gov/bill/115th-congress/house-bill/1313.

employer is only done in aggregate. The congressmen supporting H.R.1313 claimed that the bill would satisfy this exception, but as critics argued, by providing incentives for subscribers, the program undermined voluntariness. The bill was killed before a full floor vote, but that the bill even escaped its Congressional subcommittee signals a dangerous amount of support for policies that can launch genetic discrimination.

Other gaps in GINA are also troublesome and open up further vectors for abuse. To the extent that it does combat genetic discrimination, GINA only covers employment and health insurance. Life insurance has been legally denied on the basis of genetic risk for breast cancer, and in 2012 a California school forced a boy found to have a risk for cystic fibrosis in his genome to leave his class.[370] The housing market, credit companies, and other essential sectors of society have no legal obligation to respect genetic privacy.

The power of the genome means that it should not be taken lightly. Much remains to be learned about how to decode the double helix, but the trend has been one of increased discovery that undermines attempts to anonymize the data. The genome is certainly not the full story for who we are, but future findings can uncover connections we could not have perceived. The scope of this revealed information is unprecedented. An analogy of some similarity are the sperm donors before the 1990s who were told their donations would be anonymous. The technology of today, including the tests of Ancestry, one of the leaders of personal genomics, utterly undermine that pledge. Perhaps the ability to confirm one's lineage is of net benefit to society, but undoubtedly the autonomy of the donors has been compromised. The reality of today is not one they assented to.

A further risk reiterates the issues of personal data storage in general: compiled in databases, genetic data becomes vulnerable to

[370] Some more cases of genetic discrimination and the limitations of GINA are described in Zhang, "The Loopholes."

hackers and accidental disclosure. Just like behavioral data collected by data brokers and stored on their servers; like smart device data stored on the cloud, all digital data is vulnerable to cyberthreats. The danger of each of these concerns about genomic disclosure is enhanced by the state of the personal genomics industry today.

SPOTLIGHT: CAN BLOCKCHAIN SECURE THE GENOME?

Blockchain, mentioned in Chapter 3, can also be applied in the genetic realm. Innovative author and researcher David Koepsell has been working to develop a method of security for genomes called Gene-Chain that would prove substantially more difficult to hack than any current encryption technology (see Lin, "Blockchain: The Missing Link Between Genomics and Privacy?"). A blockchain storage solution would hence reduce the risk of re-identification from public sets or hacking by even determined thieves. However, it would not satisfy the pro-research activists who seek increased collaboration.

Our Genomes Should Not Be Treated "For-Profit"

Distant threats can perhaps always be imagined as the uncertainty and opacity of the future belies prediction. Nevertheless, the current state of affairs within the industry of personal genomics indicates that future imagined dangers are not merely inspired fiction: with their primary interests in profits, not patient health, the current custodians of valuable genomic data should not be trusted.

Relatively early evidence for this sprang from the focus in genomics on patenting the genome. As the Human Genome Project inched forward at the turn of the century, genomic patents were rapidly filed. As defined by the National Institutes for Health, a gene patent refers to "the exclusive rights to a specific sequence of DNA (a gene) given by a government to the individual, organization, or corporation who claims to have first identified the gene."[371] Companies could perform keen data analysis to decipher particular connections between the genome and diseases, undoubtedly of great value. However, with such patents, they could also profit from sequences pervasive across the animal kingdom; they could claim ownership of those data sequences as they appeared in billions of human beings.

Dr. Craig Venter of Celera Genomics, one of the key players in the Human Genome Project, attempted such patenting of the genome itself. Although high-profile politicians including then-President Clinton loudly voiced their dissatisfaction with these attempts, companies persisted in attempting to profit off discoveries fundamentally valuable to the human population. Filing patents for particular discoveries, researchers hoarded their findings and sought remunerations from those using the discoveries. Not merely unethical, this was also a fundamentally dangerous state of affairs, and it came to a head with the case *Association for Molecular Pathology v. Myriad Genetics*. Myriad Genetics, a biomedical company, had sequenced BRCA1 and BRCA2, genes fundamental to breast cancer development, in the mid-to-late-1990s. It had also sequenced more than forty variations of the genes, created diagnostic tests for identifying mutations, and produced a synthetic form of the gene, filing patents for all of these.

Myriad sought exclusive rights to these genes based on its research, stunting the work of competitors who promptly sued in the protracted lawsuit. At last, the case made its way to the Supreme Court in 2013. It had been through various iterations and stripped to the essentials: Are patents on human genes legal?

Unanimously, America's highest court ruled in the negative, with a few exceptions. Drawing a distinction between natural and synthetic products, the Court found that naturally occurring gene sequences could not be patented. However, synthetic products not naturally extant could

---

[371] U.S. National Library of Medicine, "Can genes be patented?"

be. Unique process-related patents, though not deliberated in the case, were also indicated to be patent-eligible.[372]

Thankfully, then, we have progressed beyond blatant attempts to claim ownership of what collectively belongs to all of us. Yet, the root problem of genetic profiteering remains prevalent. One of the first companies to offer personal genomics was 23&Me, chiefly funded by Google. When it sprang up in the first decade of the new millennium, the field was essentially an undiscovered country. It operated with abandon took advantage of this to implement suspect business practices. As state governments and eventually the FDA—tasked with consumer medical product regulation—noted, in 2013, the company promised more than it could deliver. Moreover, it identified serious disease risks without providing an adequate support structure for "diagnosed" customers.[373]

Two other sketchy practices remain broadly used. First, data continues to be treated as a corporate asset, rather than respected as fundamentally personal. For instance, deCODE genetics is a program that has ambitiously sought to gather genomic and phenotypic data of much of Iceland.[374] In 2009, the project became a private company. In 2012 as the project faced bankruptcy management decided to sell the program for $415 million to Amgen, a US biotechnology group. A few years later, the data again changed hands, acquired WuXi PharmaTech,

---

[372] For the full text of *Myriad* (2013) see this chapter's "Further Exploration" section.

[373] New York and California both warned several genetic testing startups that they required permits to operate as noted in Langreth and Herper, "States Crack Down." The FDA's warning letter can be found at https://www.fda.gov/ICECI/EnforcementActions/WarningLetters/2013/ucm376296.htm.

[374] Incidentally, the program faced strong criticism at its inception and has continued to do so. The primary reason for discomfort was the lack of choice Iceland residents had in participating. For instance, see Coghlan, "Selling the family secrets."

a Chinese firm. What began as a public project soon became a private one quite removed from the initial conditions of data collection. Fundamentally, personal genomic data was viewed as an asset for sale.[375] This is widely troubling: as one researcher, Mark Rothstein of the University of Louisville, noted from his own research, people "are not hung up on privacy so much as autonomy."[376] Participants are increasingly aware that by participating in genomics research, the genomic data they submit may not be as secure as they would hope. Nevertheless, they are willing to contribute to a cause greater than themselves. What bothers participants most is instead when companies fail to use collected data for its intended and agreed-upon purpose. Typically, such diversion has been toward corporate sale for profit.

Participant autonomy is further subverted through the secrecy with which personal genomics companies operate, coupled with the shadiness of funders. Following its public chastisement, for instance, 23&Me rebounded by reducing the strength of its marketing language. The company also signed private deals to acquire funding, while using consumer data. In 2015, the company publicly announced a partnership with fourteen private companies and universities but only listed two of their new partners and have yet to release the full list.[377] The announced partners were Pfizer and Genentech, members of Big Pharma.

Funding over 50% of biomedical studies, Pharma has a massive

[375] Amgen's purchase is described in Hirschler, "Amgen buys Icelandic gene-hunter," and WuXi PharmaTech's purchase was announced in WuXi PharmaTech, "WuXi PharmaTech Acquires NextCODE Health."

[376] Couzin-Frankel, "Trust Me, I'm a Medical Researcher," 503.

[377] Sullivan, "23andMe Has Signed 12 Other Genetic Data Partnerships."

stake in genetics research.[378] Its antibiotics, vaccines, and other drug products can be prescribed as treatment both proactively in diagnosis and after disease onset. Certainly, these products do produce great societal benefits: for instance, vaccines and drugs have nearly eradicated smallpox and polio, grave diseases that once afflicted millions. However, the suppliers have also repeatedly engaged in unethical practices. For instance, they have consistently raised drug prices after acquiring smaller companies. One high-profile case occurred in 2015 when the CEO of Turing Pharmaceuticals raised the price of Daraparin, a drug that helps those affected by HIV or other sources of weakened immune systems from $13.50 per dosage to $750. Pregnant women were particularly affected. Pharma companies have similarly abused market protections intended to stimulate innovation, employing sneaky methods such as "citizen petitions" and "pay for delay" agreements to reduce competition. Just a few years ago, for instance, pharmaceutical company GlaxoSmithKline was discovered to have marketed and sold certain drugs well beyond their FDA-approved uses.[379]

Nevertheless, these companies have strong footing as lobbyist powers. The Center for Responsive Politics compiles lists of funding expenditures by interest group and industry. For 2017, the Center estimates that more than $413 million was spent by the Health industries

---

[378] A reputable 2010 study estimated that over 60% of the more than $100 billion in biomedical research funding came from Pharma. Overall funding and the influence of Pharma have only increased since then: Stone, "Who Funds Biomedical Research?"

[379] For more on Turing Pharmaceuticals and a history of Pharma's profiteering efforts, see Baldelomar, "Where is the Line Between Ethical and Legal?" For more on the numerous ways that pharmaceutical companies reduce competition to maintain high prices and exclusive placement of their drugs, see Fox, "How Pharma Companies Game the System." GlaxoSmithKline's actions are described in *Financial Times,* "Unethical Pharma."

as a whole, with $208 million from pharma companies.[380] "OpenSecrets," the Center for Responsive Politics' website, has data running back to 1998. As the registered numbers reveal, lobbying by pharma companies steadily increased from $69 million annually. Only once annual spending reached $270 million in 2009 did it plateau.

To further provide context, according to OpenSecrets, Pharma/Health Products are by far the largest industry of lobbyists – Insurance, at $117 million is a distant second. Additionally, pharma spending costs as tabulated do not even include related categories of Health Professionals and Health Services/Health Maintenance Organizations, often partnered with pharma influencers.

With the funding they provide to genomics companies, pharmaceutical companies have a great stake in personal genomics. Yet, with their demonstrated focus on financial profits, even at the expense of consumer well-being, pharmaceutical companies cannot be reasonably expected to defend consumer privacy and security on their own. Nor can they be expected to defend these in the future with collected data. Should consumers then trust such overseers to be the custodians of genetic data? In the current ecosystem, consumers do just that. Perhaps guarantees and principles can be established to correct the obligations and direction of these parties, but until this occurs, we critically endanger our future by releasing our genetic information.

The Quantified Results Have Dubious Value

In 2005, Stanford scientist John Ioannidis sparked something of a revolution. In several papers, he lambasted the state of scientific research where contradictory and inconsistent results often propagated. One examined highly cited research studies on the effectiveness of a drug called zidovudine. The paper found that forty-five out of forty-nine research studies in prominent journals (including the *New England Journal*

---

380 The funding data, annually updated, is at OpenSecrets.org, "Pharmaceuticals/Health Products."

of Medicine, JAMA, and Lancet) that had been cited more than a thousand times, claimed the intervention was effective. Yet, only forty-four percent of subsequent studies replicated these results.[381]

More broadly, Ioannidis's August 2005 paper "Why Most Published Research Findings Are False" launched from its titular assumption, supported by several other publications at the time. Consistently, clinical research studies were found to be nonreproducible. Reproducibility is a tenet of science; after all, if a result cannot be consistently produced, then its applicability is dubious. Nevertheless, Ioannidis highlighted how isolated studies were often accepted in the research community Based on statistical trends as well as context for scores of studies, Ioannidis articulated several principles for diagnosing the truth of studies. These included that "The smaller the studies conducted in a scientific field, the less likely the research findings are to be true" and that "The hotter a scientific field (with more scientific teams involved), the less likely the research findings are to be true."[382] These papers sparked heavy traffic, and the second in particular has received more than two million views (compared to the hundreds or thousands of viewers for PLOS papers on average).[383] Nevertheless, the problems identified continue, with a 2015 paper that compiled tests of reproducibility, finding that approximately $28 billion in the United States alone is annually wasted on irreproducible preclinical research.[384]

Genetics research is part of such clinical and preclinical studies, since the efficacies of drugs influence the treatment side of personal genomics. If drugs are found to be successful at activating or suppressing essential genes in cancers or other disease, then these drugs may be

---

[381] Ioannidis, "Contradicted and Initially Stronger Effects."

[382] Ioannidis, "Why Most Published Research Findings Are False."

[383] Ibid.

[384] Freedman, et al., "The Economics of Reproducibility in Preclinical Research."

prescribed for patients. As yet, treatment recommendations by personal genomics companies for serious conditions is prohibited in the United States by the FDA, but gradually de-regulation is allowing these companies to offer more and more.

Why is it there such an epidemic of dubious results? One plausible answer points again to the influence of Pharma. As referenced in the previous section, pharmaceutical companies provide a substantial amount of funding to genomics research. Similarly, they have funded the vast majority of pre-clinical testing, and have a vested interest in having their new drugs be approved. Not coincidentally, drug tests often turn up favorably for the companies. As several commentators have argued, this may be by design.

In 2004, an editor of the *British Medical Journal* revealed a variety of tricks used by pharmaceutical companies to advocate their products. These include "Compare it with too low or too high a dose of another treatment – so the latter is less effective or has side effects" and to "Conduct your trial across a number of countries, publishing each result separately to suggest that a huge number of trials back your drug."[385] The influence of positive reinforcement appears to persist. A 2015 study evaluated 185 meta-analyses on antidepressants, where reviewers compiled studies to ostensibly provide a more thorough and unbiased perspective on the efficacy of particular drugs. Nearly eighty percent of these analyses had some sort of industry link, whether through employment, direct sponsorship, or auxiliary support (such as research grants). In itself, this was not damning but the results were: the type of reviews clearly varied by the relationship of the author to the industry. For instance, meta-analyses by industry employees were twenty-two times less likely to negatively describe a drug than others. Interestingly, this was a statistic comparable with an earlier 2006 study.[386]

[385] Law, *Big Pharma*, 45-7.

[386] Jacobson, "Many Antidepressant Studies Found Tainted."

Moreover, the results of these and other reproducibility studies suggest this is not merely an American problem; the difficulty of reproducibility is an international one, affecting research emerging from Israel and the UK among other areas.[387] Outright falsification of results may not be widespread, but some degree of questionable research practices may be more pervasive than generally acknowledged.[388]

Particularly in genomics, a controversial study from late 2017 adds further credence to this theory of self-serving studies. Researchers at Craig Venter's new startup released a paper claiming that facial structure, among other traits, can be identified from one's genome. Critics pointed to elements that evoked Ioannidis's criticisms a decade prior, with quick switches of data under discussion and small sample sizes that appeared to be homogeneous rather than appropriately representative of the population.[389] From biased data come biased results

---

[387] For instance, Baker, "First results from psychology's largest reproducibility test," and Leyser, "The dark side of research," address the global epidemic of irreproducible results.

[388] A compilation study found that up to 72% of researchers may be engaging in minor questionable research practices that do not cohere with the neutral ideals of science: Fanelli, "How Many Scientists Fabricate and Falsify Research." In her book *Big Pharma: Exposing the Global Healthcare Agenda,* Jacky Law paints a far starker picture of the pharmaceutical industry. She suggests a strong complicity of pharmaceutical representatives, drug testers, and others that feeds the production of new products that make minor and ineffective changes to old products. One section in particular, pages 62-84, excoriates the proliferation of mental health conditions, officially sanctioned in the American Psychiatric Association's bible (the DSM: Diagnostic and Statistical Manual of Mental Disorders), and largely in service of the industry. Fed by the power of placebo—the body's natural healing capabilities and mental fortitude when presented with products advocated as helpful—the myth of these products' success is perpetuated and the industry grows.

[389] Some of the fervent criticisms, which prevented its acceptance for publication at *Nature* magazine can be found at Regalado, "Does Your Genome Predict Your Face?" The original paper is Lippert, et al., "Identification of individuals by trait prediction."

and biased experimental design also leads to the same.

Alternatively, the lack of rigor in studies may be the complex product of an ecosystem that rewards rapid results and punishes slower progress. Labs and researchers that rely on funding cannot afford to have their funding dry up, and with quarterly deadlines and a bevy of competitors, rapid positive results are desired. Hence, they are incentivized to fast-track favorable studies. Even without an explicit intent to favor the products of their sponsors, researchers and reviewers may nevertheless be unwilling participants in the machine.

Another reason for mistrust of research findings is less sinister but deeper-rooted. The success of particular drugs and the correlations of multivariable analyses that aim to decipher the traits controlled by particular parts of the genome, are interpreted probabilistically. Perhaps this mirrors the truth of the world, but a probabilistic result inherently introduces uncertainty: it will not be applicable in every case.

There may also be disconnect between the public and researchers' interpretations of probability. Ideally, probabilities for disease proclivity are meant to capture some objective likelihood of individuals developing a certain disease. Via this understanding, for instance, a 70% risk of breast cancer would mean that an individual has a seventy percent risk of developing the disease.[390] Above a certain threshold, the number conveys that a patient should probably be concerned. One complication, however, emerges since diseases often progress in stages or grades. These different stages each have their own risk rates, and ultimately their own mortality rate. Different risk rates will further apply for lifestyle changes. Early onset of breast cancer may not be immediately fatal but instead have other impacts, such as impeding one's ability to focus. Certain dietary changes may help mitigate the disease's impact—vegetables and fruits high in

[390] Breast cancer includes a variety of types and is considerably more complicated than is presented here. However, it is simplified here to provide an example. Some more information on breast cancer diagnosis is at American Society of Clinical Oncology, "Biomarkers to Guide Treatment for Early-Stage Breast Cancer."

antioxidants, for instance, or Omega-3 fatty acid supplements. Adopting these changes will carry further probabilities of overcoming the disease, probabilities that extend well beyond a single number. A seventy percent chance of developing cancer is not quite as meaningful as the bare statistic appears at first glance.

In practice as well, objective truth is not readily accessible. Genomic researchers arrive at probabilities via lab testing and the numbers for risk represent a historical accounting. Cases from the past—and success rates there—dictate what numbers shape future discussion. Just like for other analyses of Big Data, then, careful consideration needs to be given to the conditions of prior experiment. Biased genomic samples representing only one ethnic group, gender, or occupation may not be reliable for a diverse population.

At an individual level this variability also applies. By virtue of living in a fluctuating environment, every individual encounters scores of events that are unique in sum. As such, particular drugs that inhibit protein receptors and particular chemicals thought to instigate cancer development may have different effects for particular individuals. Assuming there even is a seventy percent brain cancer risk that can be gleaned, whether it applies to a population on the whole or individual in particular makes a grand difference.

Ultimately, percentages may be transient myths. Quantum physics has blossomed since Albert Einstein and Robert Millikan's discovery of quantized light in the early 1900s. The field suggests that though deterministic models may make intuitive sense, they are inaccurate. At the molecular level, what's really happening involves probabilistic interactions that are unpredictable in individual scope. We expect that pressing our hand against a ball will push it away from us. In our experience, and those of everyone we know, that is the case. However, the fundamental rules are unpredictable and there exists a chance, however small, that pressing our hand against the ball will actually result in our hand passing through the spherical object, without destroying the ball itself. Biological interactions depend on the microscopic level as well. If quantum physics is to be believed—and there is good reason to do so—then genomic probabilities are merely guidelines for interaction that do not apply in every particular case. The uniqueness of the individual cannot be definitively predicted by the monolith of average statistics. Statistics may be a guide, but they are not the full story and do not dictate what must be.

None of this is to say that personal genomics is a farce or valueless. Genomic successes are numerous, and they have provided substantial insights into the body's workings. In theory, personal genomics even empowers consumers to take their health matters into their own hands. Nevertheless, sources of uncertainty from the probabilistic nature of drug testing and widespread pressure for biased design and results all suggest that the insights of personal genomics may not be as valuable as their companies insist. Through misleading diagnoses, consumers like Juliana may receive ill-founded scares and in turn be pushed towards drastic action.

In spite of all this danger, US regulators have begun a permissive trend. In 2017, led by a new FDA head with heavy pharmaceutical ties,[391] the FDA rapidly greenlit consumer products. In April, the FDA identified particular diseases and disorders that the company is now qualified to diagnose. These include schizophrenia, Parkinson's and Alzheimer's. Moreover, in the fall of 2017, the FDA reduced barriers to entry for personal genomics companies. The agency changed regulations such that personal genomics companies only need to validate one of their diagnostic tests before sending any of their products to market.[392]

---

[391] Scott Gottlieb, appointed FDA Commissioner in May 2017, has worked for years in the pharmaceutical industry and conveyed a desire for increased ability to "guide new medical innovations to market" at the expense of consumer protection: LaMotte, "Scott Gottlieb: Conflicts surround Trump's FDA pick."

[392] April's authorization is at U.S. Food and Drug Administration, "FDA allows marketing of first direct-to-consumer tests that provide genetic risk information." The FDA's more fundamental change of regulations occurred in November: U.S.

It is being made easier for profit-seekers under the guise of societal benefit to ravage consumers. The profitability focus of key parties and the unknown future mining potential of the data present strong reasons to limit engagement with personal genomics until much stronger protections are implemented.

A Case For Increased Access to Personal Genomics

Genetics may only be part of the story of who we are but it is clearly a substantial one. It is one that we have a right to access and understand. Though risks such as misdiagnosis in cases like Juliana's may exist, keys to our identity are within our DNA. By opening the market of personal genomics, we can rapidly and deservedly access valuable information and be empowered to act. Incidentally, this will also drive off substantial risks.

For a potpourri of reasons, we should open the personal genomics market. A few of these are expanded on below. First, personal genomics continues past trends of innovation and offers enormous value for both consumers and society by expanding our scientific knowledge. In turn, this bolsters consumer autonomy. Moreover, the current ecosystem of personal genomics is in fact in a strong state; it is only just that the product's value is also enjoyed by the service providers, and moreover this is necessary for sustaining future benefits. Finally, a less restrictive market will also enable a massive buy-in that will fast-track the benefits and in fact bolster privacy. By welcoming in the promise of personal genomics, rather than succumbing to imagined perils, we will activate a better future.

Food and Drug Administration, "Statement from FDA Commissioner Scott Gottlieb."

Enormous Benefits of Personal Genomics

Major revolutions in medicine have sprung from bold leaps. Without engaging in boundary-questioning research it is easy to become fixed on what has been conventionally taught and stay trapped in the mistakes of the past. This entrapment has affected all the sciences in the past, and is particularly tragic for medicine which aspires to benefit health. In medicine, improved knowledge of the body's workings enables progress. A better understanding of the body's workings allows researchers and practitioners to develop more effective treatments and moreover abandon risky and less effective ones.

Bloodletting as a treatment is a clear example of this. Through the seventeenth, eighteenth, and early nineteenth centuries, doctors widely practiced bloodletting by cuts and leeches. This was largely a holdover of an ancient and erroneous belief system. In ancient Greece, physicians operated under humourism, the belief that the body consisted of four fluids: black bile, yellow bile, phlegm, and blood; the theory held that an excess of any one, including blood, could trigger illness. Moreover, given some inevitable success cases—the body does have a remarkable ability to heal itself—many stuck steadfastly to the veracity of their convention. Hence, bloodletting found justification. As scientists understood more about the body's constituents, however, the humors could not be taken seriously. The body is composed of various fluids, but the necessity of their balance has not been empirically supported. Instead, through microscopes, the germ theory of medicine has gained ground. We can actually see microscopic entities such as bacteria that swim about. Within these pathogens we can identify genetic matter, and moreover similar genetic material for diseases of the same kind. These invaders, rather than imbalance, are now much more plausibly believed to cause disease. This simple insight, supported by repeated testing and experimentation, has led to the elimination of bloodletting as a general practice. After all, we now are aware that blood carries oxygen and nutrients around the body, which fuels energy production; while blood does replenish, it requires time; the body is typically in a weakened state when diseased as resources are focused on eliminating the invaders; hence, a loss of blood and energy productivity would further weaken the body's defenses. President George Washington, who was bloodlet after developing a fever and respiratory illness and died a few days later, was likely but one victim of the unsound practice.[393]

By developing our scientific understanding, researchers have also been able to discount prior unsound explanations motivated by prejudice. For instance, during the Black Death of the Middle Ages, hundreds of millions and perhaps 30-60% of the Eurasia population was wiped out. Ignorance mixed with prejudice led to much of Europe scapegoating Jewish communities. One aspect of this action was accurate in that Jewish communities were generally less affected than other groups; however, this was not because of deliberate poisoning of gentile wells, but rather because of certain religious practices that translated into better sanitation. By Jewish law, practitioners were compelled to wash their hands throughout the day and moreover bathe at least once a week. As we now understand, the bubonic plague or Black Death was caused by an infectious microorganism, the bacterium *Yersinia pestis*. Transmitted by rat fleas and with an incubation period up to seven days, this bacteria could in some cases be eliminated by washing one's skin.[394]

Were a similar outbreak to erupt today, we are equipped with another great defense, also indebted to good science. Antibiotics literally means "against microbial life" and refers to drugs—chemical compounds—that attack microbial invaders that could cause or are causing disease. Treatments that use herbal supplements have long been used; the ancient Egyptians, for instance, used a product moldy bread to

---

[393] A brief history of bloodletting is at Greenstone, "The history of bloodletting."

[394] Porter, "The Black Death and Persecution," and Wein, "The Black Death" discuss the scapegoating of the Jewish population during the outbreak, including stark numbers about the thousands burnt alive or otherwise killed. More information on the biology of the "Plague" can be found at *The World Health Organization's* factsheet on the disease: http://www.who.int/mediacentre/factsheets/fs267/en.

heal some infected wounds. However, a practice's true worth cannot be decoded unless its reason for working is understood. It took an accident by chemist Alexander Fleming—and keen thinking—to bring such insight. In 1928, an unwanted fungal growth on his petri dishes lay in bands separated from the *Staphylococcus* bacteria he was actually testing. Fungi was not particularly well-regarded at the time, and other researchers could easily have discarded the result as an irritating contamination. However, Fleming set aside the standard prejudice to postulate that the fungus secreted a substance that could suppress harmful bacteria. He struggled for more than a decade to produce a stable strain of the substance, and eventually Oxford researchers some twelve years later were able to do so. In 1945, through their work, penicillin began to be mass-produced as a treatment for serious diseases including pneumonia and scarlet fever. Today, pharmaceutical companies including Bayer, Merck, and Johnson & Johnson continue the work of these pioneering chemists, producing thousands of different drugs to treat disease and producing new ones. This would have been impossible without a deeper understanding that did not probe for greater information.[395]

Vaccines are another massive development, also administered chiefly by pharmaceutical companies. A vaccine contains a weakened form of an invasive pathogen that triggers production of antibodies that the body can effectively use against malevolent strains. Though vaccines have been created at least as early as 1796, when Edward Jenner combatted the deadly smallpox with injections of cowpox, a less severe disease with similar symptoms; the diseases were later discovered to share genetic similarities, and the treatment has been broadly accepted since. Through the research and funding of pharmaceutical companies, millions of people can now be immunized against influenza and other

[395] American Chemical Society "Discovery and Development of Penicillin," and Aminov, "A Brief History of the Antibiotic Era," discuss the history of antibiotics in some more depth. Clardy, et al., "The natural history of antibiotics," provides a more technical perspective.

diseases.[396] Genetic profiling technology is another revolutionary development. Through an improved understanding of earthly life's genetic composition and analysis of diverse samples, researchers have become able to predict disease proclivities. Alzheimer's and Parkinson's have been predicted from genetic content. Tay-Sachs, a degenerative disease that kills children before the age is another.[397] Nevertheless, there remains a long way to go, and initial correlations suggest there is much more that can be mined from genomic data. If genetics studies have seen dubious success rates and struggled with repeatability, it seems far less likely that the successes are mere noise than that the framework for understanding the complex relationships of gene loci to disease has yet to be uncovered. Far too many studies have uncovered relationships for all of this to be mere hogwash. By compiling stores of data, personal genomics companies make that next leap substantially more possible. Participating consumers not only receive periodic benefits for their personal health, they also contribute to the greater good, propelling our understanding of science forward like the great movements of the past.

Bolstering Consumer Choice and Equitable Profits

Besides enabling an improved understanding, personal genomics also continues a fundamental change in medicine. Arguably, the core of

[396] More information on vaccinations, also known as immunizations, is available at http://www.who.int/immunization/newsroom/events/immunization_week/2012/further_information/en. A visual history has been created by The College of Physicians of Philadelphia at https://www.historyofvaccines.org/content/history-immunization-schedule.

[397] Tay-Sachs has decisively been linked to a particular gene on chromosome 15 that codes for the Hex-A enzyme. If both copies of the gene are turned "off" then a child will have Tay-Sachs: U.S. National Library of Medicine, "Tay-Sachs Disease." No treatment is currently available, but diagnosis allows parents to prepare for the condition. Moreover, gene therapy—where genes are altered—is being explored as a possible solution for this and other diseases, as in a recent NIH request for clinical volunteers: National Institute of Neurological Disorders and Stroke, "Gene Therapy for Tay-Sachs Disease."

medicine has not changed: the Hippocratic Oath, in which doctors vow to do no harm to patients, has survived since ancient Greece.[398] However, there are several means of achieving "no harm" which produce very different sorts of treatment. For a long time, the United States and many other nations observed a paternalistic model. Doctors were regarded as the primary if not exclusive seats of knowledge. Treatments they subscribed would be accepted without question. Doctors could also withhold information from patients about diagnoses or risks if they believed the truth would be too painful for the patient.

Sometime in the mid-twentieth century this state of affairs shifted. In accordance with the UN Declaration of Human Rights and other stout pronouncements, human autonomy—a Kantian sense of dignity—was championed as fundamental. Through devastating world wars, through genocides, through widespread and systemic prejudice, the horrors of dehumanization continued to be revealed. Historical cases of abuse were also brought to the fore. Beginning in the 1930s, 399 African American men in Tuskegee, Alabama signed up for a study that ostensibly monitored their "bad blood" for free, but that in truth monitored syphilis they had contracted. For forty years, the men were only given placebos despite the fact that effective treatments, such as penicillin, were discovered in the interim. By the time the information was leaked to the public in the 1970s by a shocked employee, more than 128 of the study participants had perished from intentional misinformation, while their wives and children had gone on to contract the disease as well.[399] Another autonomy-denying action has resulted in clearer benefit. In 1951, the 31-year old Henrietta Lacks visited the Johns Hopkins Hospital for excessive bleeding. Without her knowledge, Dr. Gey collected a tissue sample from Mrs. Lacks, and these HeLa cell lines

[398] Texts of the classical Hippocratic Oath and a modern version can be found at Tyson, "The Hippocratic Oath Today," along with commentary by doctors and others.

[399] Nix, "Tuskegee Experiment."

from her cervical cancer would be used to develop polio, and launch serious attempts at cloning, among other applications; they continue to be used in cancer research, although descendants of Mrs. Lacks now sit on the advisory board that approves cell use.[400]

Based on these and other incidents of sketchy behavior, the United States Congress issued the 1974 National Research Act which in turn produced the Belmont Report in 1978. This influential document continues to guide medical studies that involve human patients and established three core principles to be respected: (1) Respect for persons, (2) Beneficence, and (3) Justice. An explicit offshoot of the first principle identified by the document is "informed consent." Study participants must, "to the degree that they are capable, be given the opportunity to choose what shall or shall not happen to them."[401] These are fundamentally affirmations of patient autonomy and other nations are also upholding the value. For instance, the Medical Protection Society, the world's leading protection organization for doctors, dentists, and other medical professionals boasts 300,000 members in the UK, South Africa, Singapore, and other nations. As the organization acknowledges, though specific valuations of consent vary across the globe, "the concept of 'informed consent' is now the most common standard" for medical procedures.[402]

[400] More information about Henrietta Lacks's story can be found at Johns Hopkins Medicine, "The Legacy of Henrietta Lacks." *The Immortal Life of Henrietta Lacks* by Rebecca Skloot dives into the events in greater depth and intersperses discussion of related bioethical issues.

[401] The Office for Human Research Protections through the US Department of Health and Human Services provides a hub for information on ethical processes and more for human research in the United States: https://www.hhs.gov/ohrp. It links to the full text of the Belmont Report as well as upcoming educational events and regulation changes.

[402] Medical Protection, "Content to consent?"

More recent developments in the United States have also forwarded this push towards patient control of data. The landmark HIPAA of 1996 was created "to improve portability and continuity of health insurance coverage" as well as "to improve access to long-term care services and coverage" and "simplify the administration of health insurance" among other purposes.[403] All of these purposes work towards transparency and easy accessibility of records for patients, which is in turn valuable to empower patient autonomy. Subsequent amendments have continued this trend: the Privacy and Security Rules were enacted in the early 2000s to bolster cybersecurity standards for electronic health records and moreover require patient consent before sharing of the data; HITECH in 2009 was intended to promote the adoption and meaningful use of health information technology—in turn, this too would improve accessibility by patients to enable their decision-making. GINA, as has been discussed earlier, also sought to encourage patient gene sequencing by allaying fears about future discrimination.

Personal genomics now offers a further boost for patient autonomy. Through these sequencing programs, patients acquire insight into their genomes. Undoubtedly, misdiagnoses are possible, but they are a risk in any field. Patients should be trusted and empowered to make the best decisions for themselves. Enabling access to personal genomics in an open market allows just this. Not only are citizens equipped to evaluate information themselves, but encouraging market growth will increase consumer choice.

With their years of study and training, doctors are certainly still worthy of respect and deference. However, the proliferation of today's conditions and rapid pace of research mean that even they cannot possibly stay abreast of all developments. With today's information age, patients are more equipped than ever to help themselves. The explosion of reputable information available through the web allows patients to

---

[403] Links to the text of HIPAA as well as later developments can be found at U.S. Department of Health and Human Services, "HIPAA for Professionals."

research and reason themselves. For those overwhelmed and wary of the abundance, traditional recourses of medical professionals remain available. However, this should not be the mandatory default; those eager to venture out and conduct research themselves should also have this option available. This is the natural way to continue the powerful and essential trend of empowerment.

One remaining concern with the current ecosystem for autonomy is that personal genomics companies will mine consumer data in unintended ways. Through the convoluted FDA regulations enacted in the present, these companies are arguably not providing a health service and hence are not subject to HIPAA. Hence, sharing the data of consumers without consent is not necessarily prohibited. However, several reasons make these not nearly as dire a situation as some might claim. First, United States and global companies must reveal their data sharing policies in privacy policies. Many of the high-profile companies including 23&Me explicitly state that they will not share the data; they may conduct further testing upon their databases of consumer genomes and submitted phenotypic data, but this will remain in-house. Even for companies that do not make such promises, strong protections do remain in place from GINA against genetic discrimination in arguably the most important fields of employment and health insurance. Relatedly, with an open market, consumer choice will be bolstered so that companies with amenable privacy policies thrive.

Moreover, as has been noted, consumers who supply genetic data do stand a risk of re-identification. However, as this is becoming acknowledged by the industry, consumers continue to flock to the services. One global venture, recognizing the unlikelihood of anonymization, makes no guarantees about the privacy of participants. Instead, the Personal Genome Project has begun the revolutionary "open consent" framework that requires prospective patients to pass a holistic exam testing the ramifications of re-identification and other participation risks. Even though it highly publicizes serious risks, enrollment continues. Since the program was begun in 2010 by Harvard

Professor George Church, it has made consistent progress: as of April 2017, it has 5,500 worldwide. The not-for-profit project allows participants to access their own data, and ultimately aspires to thrust genomic research forward.[404] Are consumers daredevils; are their rational capabilities stunted? No: rather, as has been repeatedly supported, consumers recognize the risks but are even more captivated by the hope of advancing genetic understanding. Some of these participants have faced degenerative diseases in family or friends; some are simply benevolent and wish to improve the lives of those less fortunate. In any case, as one researcher, Mark Rothstein of the University of Louisville, has noted, people "are not hung up on privacy so much as autonomy."[405] What troubles participants most is failing to use the data for its intended and agreed-upon purpose of research, such as a purely for-profit use. Clarity of data's purpose through privacy policies and other means can rectify this flaw where it appears, if it does emerge. Consumer autonomy should be valued.

Evaluating the Current Landscape

Much of the data brought up by the critic above is accurate; however, the interpretations are too limited. In fact, the personal genomics industry is a beneficial one in a healthy state. The fearmongering is overstated as a more perceptive dive into the funding entities reveals. Additionally, the issue of reproducibility that the research faces is a natural result of science moving forward rather than becoming perverted.

Certainly, for biomedical research in general, and personal genomics in particular, there is a massive amount of funding by pharmaceutical companies. Yet, this is not a bad thing. For one, despite

---

[404] More on the Personal Genome Project can be found at Ball, et al., "Harvard Personal Genome Project," and the informational page for the enrollment site is http://www.personalgenomes.org/us.

[405] Couzin-Frankel, "Trust Me, I'm a Medical Researcher," 503.

its many detractors, Pharma performs an extremely valuable service. The drugs the industry produces have great societal impact. Moreover, the apparently unethical behavior can be normalized by considering two further factors. First, these companies do invest in substantial efforts to lower costs and make new drugs available. Secondly, the capitalistic nature of the pharmaceutical market requires that the companies operate as they do.

With regard to price-reducing efforts, many pharmaceutical companies invest in price-reducing initiatives. For instance, the Partnership for Prescription Assistance (PPA) is a free initiative sponsored in part by drug companies working with numerous American health organizations. Through PPA, patients can acquire free or low-cost medications. Drug companies also have their own beneficial programs, and have begun group initiatives. In 2012, for instance, thirteen high-profile drug companies committed to donating drugs and conducting research to control if not eliminate ten neglected tropical diseases by 2020. These companies, including Pfizer, Johnson & Johnson, Merck, and Novartis, pledged $785 million and appear to be on track to the goal. Many other programs have commenced that aspire to eradicate dangerous disease.[406]

Moreover, pharmaceutical companies are products of their environment. The research they conduct is extremely expensive: acquiring the chemicals, mixing them in the right cocktails, engineering vaccines, and certifying their products is a labor and cost-intensive process. 2012 estimates, for instance, pegged the development of a new drug at four to eleven billion dollars.[407] It only seems fair that these

[406] For the 2012 initiative see Kelland, "Big Pharma donates drugs." The "Access to Medicine" index, published every two years, tracks efforts by pharmaceutical companies to make valuable drugs accessible to all. These efforts include donations and cost subsidies by the pharma companies, with the latest from 2016: https://accesstomedicineindex.org.

[407] Per 2012 estimate: Herper, "The Truly Staggering Cost of Inventing New Drugs."

companies, producing an enormous societal benefit receive some compensation. As discussed in Chapter 8, performing a good service for profit and using consumer data for this end need not be a disrespectful service. Even beyond this, these companies require some profit merely to survive. If the money does not come from customers and their data, where else would it come from? The options are limited and unpalatable. Perhaps more governments could follow the socialized medicine system of Scandinavia and Canada. However, this incurs a greater cost on the population-at-large: the majority of citizens subsidize the costs incurred by a few through taxes and reduced alternative programs by government. And certainly, the companies cannot possibly operate *pro bono* to survive and produce the valuable drugs of tomorrow.

The inefficiency of Pharma testing indicates further that the companies are not exclusively interested in profiteering. It takes billions to develop a new drug and development costs are increasing, while fewer drugs are being approved by the FDA than before. Returns from R&D have dropped from around 10% to 4% and less recently, resulting in substantially reduced profit margins.[408] Though some critics point to the massive amounts of money (tens of billions by the large companies) that these pharma companies spend on marketing, development costs and the particular struggles of large firms against younger competitors with more data-driven models suggest that this spending may be warranted; it would not have much value in research and only contribute further to waste. Perhaps greater efficiency can be achieved here to reduce the costs incurred, but to this effect, new insights from personal genomics databases will be of help.

For those who criticize how pharma companies "game" the system by constructing experiments, they should also note that struggles for progress are neither new nor unusual. In the past, studies tended to

[408] Deloitte has tracked the profitability of the pharmaceutical industry since 2009 in their annual reports. The 2017 report is at Deloitte, "A New Future for R&D." For alternatively sourced stats, see Sarich, "Pharma Companies Spend 19x more on Marketing."

confirm false paradigms because of ingrained teachings. For instance, the anti-bloodletting movement encountered resistance even through the twentieth century. In a widespread sentiment that set back science, researchers asserted that they were not prepared to discard therapies "validated by both tradition and their own experience on account of somebody else's numbers."[409] Armed with the "wisdom" of their textbooks and only faced with a world where bloodletting was common practice, doctors had little empirical or theoretical reason to act otherwise; in fact, they were conditioned to reinforce the erroneous establishment.

The populated sphere of competing studies with unclear results is in fact a positive one. It means that science is working. Reproducibility is the sign of a strong scientific theory and the failure of reproducibility only indicates that continued research is needed. The onus falls on researchers to continue best practices of science: they must maintain skepticism, challenge conventional thinking, and evaluate the statistical basis for studies. The solutions Dr. Ioannidis and others have recommended to combat the uncertain state of current studies depend on the vigilance of researchers.

In the current ecosystem, pharmaceutical companies are right to seek profits. If something must be changed, then the budgets of governments must be seriously considered, or inefficiencies explored more deeply.

To achieve an enlightened state and produce still other benefits, one other substantial change to the current ecosystem may be worthwhile: making the fruits of genetic research more broadly accessible. The work being done by biomedical companies in research uncovered valuable information, and it should be used for benefit, not stored away for isolated profit.

---

[409] See Rangachari, "Evidence-based medicine," 282.

SPOTLIGHT: What Else Can Genetics Do?

CRISPR is a gene-editing technology that has been frequently featured in the news recently. Taken from a particular bacterial immune system and used in conjunction with the Cas9 protein, CRISPR is special as it allows for faster and easier gene editing than ever before.

Since its development in 2012 by UC Berkeley scientists, CRISPR has repeatedly produced breakthroughs in the abilities of scientists to "play God." One of the most recent allows scientists to correct a genetic mutation responsible for the heart condition hypertrophic cardiomyopathy. A disorder that causes thickening of heart ventricles, HT constricts blood flow and can lead to death. Through researchers' efforts at Oregon State University, however, CRISPR was shown to be successful at removing the mutation (Barclay, "Scientists Successfully Used CRISPR to Fix a Mutation").

Gene editing has been controversial for a long time, and some feel that humans should not be interfering so deeply with our natural proclivities. In particular, some are wary that editing will lead individuals to produce designer babies with desired physical traits, removing chance from the reproductive process, and eventually perhaps even designing for particular personality traits. On an environmental scale, this could be devastating for genetic diversity, meaning that a single disease could well be more likely to destroy the human race. Implications for children who did not meet specifications are also severe, as embryos that do not meet specifications could be terminated by zealous parents-to-be. The societal implications are vast as well, with individuals who have "disabilities" likely to be further removed from the rest of society, pitied rather than appreciated. With its goods of precluding debilitating disease and its dangers of abuse, CRISPR is—like many of the technologies discussed in this book—a double-edged blade. It's just a blade that works at the genetic level.

The Necessity of Massive Buy-In

Directly and indirectly, a massive buy-in to personal genomics will produce substantial benefits. Encouraging rather than restricting personal genomics is hence the way to go for several reasons.

First, by accepting personal genomics as legal, citizens will not have to turn to illegal and dubious alternatives. Prohibition of controversial items or procedures in the past—alcohol in the early 1900s, for instance—has only pushed these underground and encouraged more dangerous activities, such that mortality rates exploded.[410]

Moreover, in the quest for understanding, speed is an ally. The sooner investment and engagement occurs, the sooner the results will come through. As with a Big Data project, representative data is needed. Representative genetic data requires sequences from diverse races. As sequencing projects like the Personal Genome Project have noted, genetic diversity will also require sheer quantity to decipher.[411] Only with a high volume can meaningful results be separated from coincidental relationships within a data set.

Beyond data collection of humans, there are not currently viable alternatives. Animal experimentation has been pursued as one possibility. After all, we share DNA with nearly all of life. It is estimated, for example, that mice and men share 97.5 percent of the same DNA,[412] while humans and chimpanzees share 98.8 percent.[413] We also share about 60 percent of our DNA with the common fruit fly, the same percentage (though different shared genes) with bananas, and more than

---

[410] See Thornton, "Alcohol Prohibition Was a Failure" for one sharp analysis on the dangers of prohibition.

[411] The Personal Genome Project, for instance, has set an enrollment goal of 100,000 participants: Duncan, "On a Mission to Sequence the Genomes of 100,000 People."

[412] Coghlan, "Just 2.5% of DNA Turns Mice into Men."

[413] American Museum of Natural History, "DNA: Comparing Humans and Chimps."

6 percent similarity with some bacterial strains.[414] However the miracle of life originated, it is clear that much of the genome has been preserved among the living and passed on. Continuing to probe corresponding genes and segments in other creatures can help us understand connections of interest in human beings without having to use human beings. Similarly, testing on synthetic creatures is another possibility. The discovery of stem cells has meant that researchers can manipulate somatic cells to produce tissues (groups of cells) and organs (groups of tissues). For DNA testing, the advancement of CRISPR allowed researchers to produce synthetic life without a donor.

While intriguing options, neither is a serious alternative for human sampling and testing. At an ethical level, so long as the basis for "personhood" remains unclear, we may be violating the rights that forms of life besides *Homo sapiens* possess. Nonhuman creatures cannot provide the same sort of consent that articulate reasoning creatures can. Moreover, tests have not transferred well from nonhumans to humans. Various differences such as internal biome conditions (such as with the bacteria typically in a human gut) and traditional environments can affect the percentage risks that genes and treatments have from species to species.[415] Ultimately, human participation appears necessary for us to access the keys of the genome, and to this end, more data samples will best provide more reliable results.

The benefit of a massive buy-in is bound to have a snowballing effect for improvement. Through data analysis and improved knowledge of correlations, more focused wet-lab testing can be performed. This would reduce the cost and length of time required to perform testing in order to produce viable drugs. Certainly, this could reduce wasteful spending by pharmaceutical companies and institutions, hence allowing

---

[414] Trost, et al., "Comparing the Similarity of Different Groups of Bacteria to the Human Proteome."

[415] For instance, see Harris, "Drugs That Work in Mice Often Fail," and Akhtar, "The Flaws and Human Harms of Animal Exploitation."

their research to be more effective. In turn, this would allow drug companies to price their products at more affordable rates. Embracing this possibility, 23&Me has already begun an initiative for drug companies to investigate its database. Data is de-identified as best as is possible given current knowledge, and consumers explicitly sign a waiver to participate. 80% of the company's customers are currently participating and forward-thinking pharmaceutical companies have invested hundreds of millions of dollars towards the effort as of September 2017.[416]

Finally, a massive buy-in also affords an important privacy protection. Much as with reputation protection in the "right to be forgotten," by wide exposure, participants and the population are equally protected. If everyone's data is equally vulnerable, there is substantially less room for blackmail and data abuse. While all are exposed, all are also equally protected.

Powerful technologies naturally spark powerful reactions. Critics, perhaps influenced by that pervasive bias for the familiar past, are quick to jump on projected catastrophe if innovations are adopted. Yet, these impulses bury the truth and stymie development. Fears of information overload were vocalized against the printing press; one prominent critic claimed propagation of books would overwhelm the human mind and in turn result serious harm. Similarly, critics renewed this same refrain against radio, television, the Internet, and smartphones, while launching other critiques such as the danger of empowering an uneducated polity to mob rule and censorship. [417] The truth has been that these modes of technology have enriched our lives. The printing press sparked dialogue, motivating freer expression and propelling a richer marketplace of ideas. Ultimately, it and its future descendants have become tools to democratize information and learning, and promote

---

[416] Feldman, "23andMe's Next Phase."

[417] On the printing press, see Blair, "Reading Strategies for Coping with Information Overload," and on other technologies, see Bell, "Don't Touch That Dial!"

justice. Dissidents and terrorists may also employ these technologies to spread hateful ideas and motivate senseless violence, but the openness of the market and power of transparency have permitted truth to prevail. The great value of the printing press has been rightfully taken on by society, triumphing over the possible dangers.

For all of the reasons given above, personal genomics possesses enormous value. It holds great potential to advance our understanding of science which will in turn translate into improved treatments. Moreover, like the printing press, it empowers the individual, and is in fact situated to be used for good in the present ecosystem. With encouragement and a massive buy-in, the benefits will be set to arrive faster. Like the printing press, the benefits of the technology will triumph over the risks; we must trust in the power of citizens and push forward for a better tomorrow.

In Conclusion

Personal genomics offers great value in providing consumers potential keys to their health and hence empowering them. With broad buy-in, new insights can be expedited and consumer privacy potentially protected. Moreover, like previous developments in science, unforeseen benefits are bound to emerge from faster research and development. The insights gleaned, however, come at some cost with the risk of data breach as well as data use unintended by consumers that causes harm.

Whatever our personal evaluations of risk, the current regime appears inadequate. It is either a stepping stone towards increased restriction or increased market freedom. Juliana's case is a moderate exaggeration at the present, but a plausible future. What is allowed, what ought to be allowed, and how the benefits of genomics should be leveraged remain questions of the utmost importance. Given the significance of our genomes—to the identity of ourselves and our loved ones—these must be swiftly addressed by each of us as we decide how to move forward.

Further Exploration

Curated by British personal genomics company DNA Worldwide, a brief timeline of "The History of DNA" is available at **https://www.dna-worldwide.com/resource/160/history-dna-timeline**. Compiled by the journal *Nature*, more technical timelines are at **https://www.nature.com/milestones/miledna/timeline.html**.

The thirty-six members of the intergovernmental economic cooperation Organization for Economic Co-Operation and Development (OECD) still abide by privacy guidelines from 1980. The organization continues to modify these as it deems necessary. For the latest updates pertaining to biotechnology, see the corresponding web page at **http://www.oecd.org/sti/biotech**.

The ***Genomics Law Report*** is an online publication by reputed corporate law firm Robinson, Bradshaw, and Hinson, which is based out of the Carolinas. Active since 2009, the site features posts by the law firm's affiliates and contains news on current developments related to genomic law, including patent developments as well as international, national, and local rulings. Posts on the site also include more links to follow-up materials: **https://www.genomicslawreport.com/index.php/category/general-interest/privacy-general-interest**.

Supported by the federal **Department of Health and Human Services** in their Office for Civil Rights, the **OCR portal** lists all breaches of health data that affect five hundred or more individuals. This follows from passage of the 2009 HITECH (Health Information Technology for Economic and Clinical Health) Act: **https://ocrportal.hhs.gov/ocr/breach/breach_report.jsf**.

The **Personal Genome Project** was begun in 2014 by Harvard Professor George Church, with more than fifty-five hundred volunteers as of April 2017. It seeks to completely sequence and publish the genomes and medical records of one hundred thousand volunteers to forward research into personal genomics and medicine. It includes branches in the United States, Canada, United Kingdom, and Austria: **http://www.personalgenomes.org**.

Genetic profiling companies are increasing in number across the globe. One sample profiling company is **Complete Genomics**: **http://www.completegenomics.com**.

Directed by Dr. Latanya Sweeney, the **Data Privacy Lab** seeks to "provide intellectual leadership to society in shaping the evolving relationship between technology and the legal right to or public expectation of privacy in the collection and sharing of data." As such, it includes a compendium of published papers and articles on numerous aspects of privacy, as well as courses and contact information for professionals willing to discuss issues in more depth. The topics explored include re-identification, suggestions for avoiding e-mail spam, and suggestions on secure means of sharing sensitive data. Genomic privacy is also highlighted, with sections describing the means of achieving HIPAA certification as well as risks and benefits of genomics development: **https://dataprivacylab.org**.

The United States' **NIH Privacy Portal** features archives as well as the latest released news on all things genomic: **https://www.genome.gov/27561246/privacy-in-genomics**.

The full text of **2008's** landmark **GINA**, along with some supporting explanatory documents can be found at **https://www.hhs.gov/hipaa/for-professionals/special-topics/genetic-information/index.html**.

Conclusion

We find ourselves today amidst the wild lands of morality and technology. Some fear that the rise of artificial intelligence and other technologies will steal our jobs, then transform tomorrow into an uninhabitable blasted heath. Others read a rose-colored future from the tea leaves, a utopia to rival Edenic splendor. Perhaps you see light in between. Whatever may be realized, we have the opportunity now to decisively shape what comes next. Our decisions will affect each of us and each of our fellow citizens of the world.

Echoed behind our deliberations, the whisper of "Rosebud" lives on. Its legacy as an intimate curiosity resonates all the more in today's digital society, where our actions are documented and preserved. Whether we wish to safeguard or destroy our personal rosebuds, whether we wish to broadcast them for all the world to see, each of us will have to decide. To perform such reflections, we will also have to meditate upon the value of privacy as well as other rights we hold dear.

Today, the digital landscape harbors great benefits and dangers, both yet to be fully mined. Now is the opportunity to build the future of our choosing, and to do so we must engage in open discussion. We will need to involve all stakeholders—established professionals with technical expertise, government officials vested in power, and consumers young and old who will be the most affected. Corporations, recluses, social animals, and other individuals, all are stakeholders. We will need to listen to make our own voices heard. In privacy matters, as with all else, a stronger future will require impassioned collaboration.

Acknowledgements

A great many people have helped to make this book a reality, and I would like to highlight a few who have had particularly great influence. I am extremely grateful to my former mentors and teachers, especially Mr. Ed McAdams, Mrs. Julie McKee, Mrs. Koppeis-Bowles, Mr. Brent Concilio, Mrs. Barbara Palmer, and Mr. Greg Tucker who fostered in me a love for writing and deeper inquiry. More recently, Professor Michael Kerbel has helped to feed and deepen my appreciation for cinema, and to use it as a lens for all things. Through their own example and through their guidance, Professor Silverman, Professor Matthew Croassmun, and Professor Kenneth Winkler have helped me to appreciate the value of, and to wholeheartedly engage with, diverse perspectives. Professor Dr. Alex Pentland of MIT's Media Lab has also enormously expanded my technical grasp of Big Data, opening my eyes to the bounties of strong data analysis.

Since becoming a certified privacy professional, I have additionally benefited from the IAPP which has offered an engaging forum to contribute and continue learning about privacy issues across the globe. A great debt is also owed to all of the other authors referenced in this work for their own research and provocative pieces. Their keen and diverse insights have motivated me; I hope you will visit many of their works in the References section that follows.

Several individuals have provided tremendous feedback during the course of the manuscript's completion. Thank you JM for your comprehensive suggestions. Thank you also to MDC and MK for your perceptive and patient feedback. And thank you, DM, for your insights, particularly on aviation.

Last but certainly not least, thank you Mom and Dad for your unceasing encouragement.

References

"10 Early ISPs and What Has Become of Them." *Internet Service Providers.* March 7, 2011. http://www.internetserviceproviders.org/blog/2011/10-early-isps-and-what-has-become-of-them.

"23&Me." *23&Me.* Accessed July 7, 2017. https://www.23andme.com.

"23andMe, Inc. 11/22/13." *Food and Drug Administration,* November 22, 2013. https://www.fda.gov/ICECI/EnforcementActions/WarningLetters/2013/ucm376296.htm.

"About IGSR and the 1000 Genomes Project." *IGSR: The International Genome Sample Resource.* Accessed December 13, 2017. http://www.internationalgenome.org/about.

Accenture. "Igniting Growth in Consumer Technology." Communications, Media, & Technology, Accenture, Accenture. 2016. https://www.accenture.com/t20160108T124537__w__/us-en/_acnmedia/PDF-3/Accenture-Igniting-Growth-In-Consumer-Technology.pdf.

"Access to Medicine Index." *Access to Medicine Foundation.* Accessed December 13, 2017. https://accesstomedicineindex.org.

Ackerman, Spencer. "Microsoft, Facebook, Google and Yahoo Release US Surveillance Requests." *The Guardian,* February 3, 2014. https://www.theguardian.com/world/2014/feb/03/microsoft-facebook-google-yahoo-fisa-surveillance-requests.

ACLU Northern California. "Location-Based Services: Time for a Privacy Check-In." ACLU Northern California. Accessed July 2, 2017. https://www.aclunc.org/sites/default/files/asset_upload_file183_9627.pdf.

ACLU. "ACLU v. NSA—Challenge to Warrantless Wiretapping." September 10, 2014. https://www.aclu.org/cases/aclu-v-nsa-challenge-warrantless-wiretapping.

______. "Doe v. Mattis – Challenge to Detention of American by U.S. Military Abroad." Last Updated April 7, 2018. https://www.aclu.org/cases/doe-v-mattis-challenge-detention-american-us-military-abroad

______. "What Is Net Neutrality." Last Updated December 2017. https://www.aclu.org/issues/free-speech/internet-speech/what-net-neutrality.

Acosta, Jim, and Jeremy Diamond. "Obama ISIS Fight Request Sent to Congress." *CNN,* February 12. http://www.cnn.com/2015/02/11/politics/isis-aumf-white-house-congress.

Adams, Douglas. "How to Stop Worrying and Love the Internet." Reprinted from *The Sunday Times,* Aug. 22, 1999. http://www.douglasadams.com/dna/19990901-00-a.html.

Adhikari, Kaustubh, et al. "A genome-wide association study identifies multiple loci for variation in human ear morphology." *Nature Communications* (June 2015): 1-10. doi: 10.1038/ncomms8500

AirForce Technology. "Hummingbird Nano Air Vehicle (NAV)." Accessed May 1, 2018. https://www.airforce-technology.com/projects/hummingbird-nano-air-vehicle.

Ajunwa, Ifeoma. "There's No Guarantee of Anonymity." *New York Times*, March 4, 2015. https://www.nytimes.com/roomfordebate/2015/03/02/23andme-and-the-promise-of-anonymous-genetic-testing-10/theres-no-guarantee-of-anonymity.

Akhtar, Aysha. "The Flaws and Human Harms of Animal Exploitation." *Cambridge Quarterly of Healthcare Ethics* (October 2015): 407-19. doi: 10.1017/S0963180115000079.

Alderman, Ellen, and Caroline Kennedy. *The Right to Privacy*. New York: Vintage Books, 1997.

Amatriain, Xavier, and Justin Basilico. "Netflix Recommendations: Beyond the 5 stars (Part 1)." *Netflix Tech Blog*, April 5, 2012. https://medium.com/netflix-techblog/netflix-recommendations-beyond-the-5-stars-part-1-55838468f429.

American Chemical Society International Historic Chemical Landmarks. "Discovery and Development of Penicillin." Accessed December 13, 2017. https://www.acs.org/content/acs/en/education/whatischemistry/landmarks/flemingpenicillin.html.

American Museum of Natural History. "DNA: Comparing Humans and Chimps." Accessed July 7, 2017. http://www.amnh.org/exhibitions/permanent-exhibitions/human-origins-and-cultural-halls/anne-and-bernard-spitzer-hall-of-human-origins/understanding-our-past/dna-comparing-humans-and-chimps.

American Society of Clinical Oncology. "Biomarkers to Guide Treatment for Early-Stage Breast Cancer." July 10, 2017. https://www.cancer.net/research-and-advocacy/asco-care-and-treatment-recommendations-patients/biomarkers-guide-treatment-early-stage-breast-cancer.

Aminov, Rustam I. "A Brief History of the Antibiotic Era: Lessons Learned and Challenges for the Future." *Frontiers in Microbiology* (December 2010). doi: 10.3389/fmicb.2010.00134

Anderson, Nick, and Ann E. Marimow. "Former D.C. nightclub owner Antoine Jones sentenced on drug charge." *Washington Post,* May 1, 2013. https://www.washingtonpost.com/local/antoine-jones-pleads-guilty-to-drug-charge/2013/05/01/1109c268-b274-11e2-bbf2-a6f9e9d79e19_story.html?utm_term=.6c95698575d8.

Angwin, Julia. "Meet the Online Tracking Device That Is Virtually Impossible to Block." *ProPublica,* July 21, 2014. https://www.propublica.org/article/meet-the-online-tracking-device-that-is-virtually-impossible-to-block.

______. "The Web's New Gold Mine: Your Secrets." *Wall Street Journal,* July 30, 2010. https://www.wsj.com/articles/SB10001424052748703940904575395073512989404.

Annas, George J., and Sherman Elias. *Genomic Messages: How the Evolving Science of Genetics Affects Our Health, Families, and Future.* New York: HarperOne, 2015.

Annunziata, Marco. "The Value of interconnectedness." General Electric, June 22, 2015. https://www.smartindustry.nl/site/assets/files/1414/si-wp-ge-valueofinterconnectedness.pdf.

Appleby, Tyra. "What is Phishing? InfoSec Institute. Accessed September 18, 2017. http://resources.infosecinstitute.com/category/enterprise/phishing.

Apps, Peter. "DDoS Cyber Attacks Get Bigger, Smarter, More Damaging." *Reuters*, March 5, 2014. http://www.reuters.com/article/us-cyber-ddos-idUSBREA240XZ20140305.

Asaro, Peter. "Dr. Peter Asaro on Drone Technology in Eye in the Sky." *Science and Film*, April 19, 2016. http://scienceandfilm.org/articles/2686/dr-peter-asaro-on-drone-technology-in-eye-in-the-sky.

Aspa, Jocelyn. "Top Cloud Security Companies." *Cloud Investing News*, November 7, 2017. https://investingnews.com/daily/tech-investing/cybersecurity-investing/top-cloud-security-companies.

Association for Molecular Pathology v. Myriad Genetics, Inc., 569 U.S. 12-398 (2013).

Associated Press. "FCC Prepares Rules for the Post-Copper Phone World." August 5, 2015. http://www.cbsnews.com/news/fcc-prepares-rules-for-the-post-copper-phone-world.

Atherton, Kelsey D. "How It Works: The Thermal Camera That Found the Boston Bomber." *Popular Science*, April 25, 2013. http://www.popsci.com/technology/article/2013-04/how-works-awesome-thermal-camera-found-boston-bomber.

Bailyn, Bernard. *The New England Merchants in the Seventeenth Century*. Rev. ed. Cambridge: Harvard University Press, 1979.

Baker, Monya. "First results from psychology's largest reproducibility test." *Nature,* April 30, 2015. doi: 10.1038/nature.2015.17433.

Baldelomar, Raquel. "Where is the Line Between Ethical and Legal?" *Forbes,* July 21, 2016. https://www.forbes.com/sites/raquelbaldelomar/2016/07/21/where-is-the-line-between-what-is-ethical-and-legal.

Ball, Madeleine P., et al. "Harvard Personal Genome Project: lessons from participatory public research." *Genome Medicine* (February 2014). doi: 10.1186/gm527.

Barbash, Fred. "'Tesla-in-a-tunnel': Elon Musk to build 125 mph underground transit line serving Chicago O'Hare." *Washington Post*, June 14, 2018. https://www.washingtonpost.com/news/morning-mix/wp/2018/06/14/tesla-in-a-tunnel-elon-musk-to-build-125-mph-underground-transit-line-from-ohare.

Barclay, Eliza. "Scientists Successfully Used CRISPR to Fix a Mutation that Causes Disease. This Is Huge." *Vox*, August 2, 2017. https://www.vox.com/science-and-health/2017/8/2/16083300/crispr-heart-disease.

Barrett, Brian. "Amazon's New Unlimited Cloud Storage is Absurdly Cheap." *Wired*, March 28, 2015. https://www.wired.com/2015/03/amazon-unlimited-everything-cloud-storage.

Barrett, David. "Burglars Use Drone Helicopters to Target Homes." *Telegraph*, May 18, 2015. http://www.telegraph.co.uk/news/uknews/crime/11613568/Burglars-use-drone-helicopters-to-identify-targe-homes.html.

Barry, John M. *Roger Williams and the Creation of the American Soul.* New York: Penguin, 2012.

Barth-Jones, Daniel C. "The 'Re-identification' of Governor William Weld's Medical Information: A Critical Re-examination of Health Data Identification Risks and Privacy Protections, Then and Now." *Future of Privacy Forum Working Paper*, June 18, 2012. https://fpf.org/wp-content/uploads/The-Re-identification-of-Governor-Welds-Medical-Information-Daniel-Barth-Jones.pdf.

Bates, Karl. "Identical Twins, Different Lives." *Duke University*, January 12, 2016. https://research.duke.edu/identical-twins-different-lives.

BBC. "Google's Giant Penalty: The Internet Reacts." June 27, 2017. http://www.bbc.com/news/technology-40418197.

Bell, Vaughan. "Don't Touch That Dial!" *Slate*, February 15, 2010. http://www.slate.com/articles/health_and_science/science/2010/02/dont_touch_that_dial.html.

Bergen, Kathy. "Chicago seeking 'smart-city' tech solutions to improve city life." *Chicago Tribune*, April 29, 2016. http://www.chicagotribune.com/news/globalcity/ct-chicago-smart-city-global-city-20160421-story.html.

Bernstein, Bobby. "Top 5 Best Auto-Follow Drones." *Heavy.com,* May 19, 2016. http://heavy.com/tech/2016/05/top-best-auto-follow-drones-autonomous-drone-phantom-4-ghostdrone-airdog.

Bird, Mike. "Why Europe Hates American Tech Companies." *Business Insider,* April 2, 2015. http://www.businessinsider.com/why-does-europe-hate-tech-companies-2015-4.

Blair, Ann. "Reading Strategies for Coping with Information Overload ca. 1550-1700." *Journal of the History of Ideas* (January 2003): 11-28. doi: 10.2307/3654293.

Blake, Thomas, Nosko, Chris, and Steven Tadelis. "Consumer Heterogeneity and Paid Search Effectiveness: A Large-Scale Field Experiment." *Econometrica: Journal of the Econometric Society* 83 (1) (2015): 155–174.

Blockchain. "Market Price (USD)." Accessed July 6, 2017. https://blockchain.info/charts/market-price?timespan=all.

Boffey, Daniel. "Google 'faces €1bn-plus fine' from EU over market dominance." *The Guardian,* June 16, 2017. https://www.theguardian.com/technology/2017/jun/16/google-fine-eu-case-shopping-search-result.

Bogart, Michele Helene. *Artists, Advertising, and the Borders of Art.* Chicago: University Of Chicago Press, 1999.

Bogost, Ian. "The Internet of Things You Don't Really Need." *The Atlantic,* June 23, 2015. https://www.theatlantic.com/technology/archive/2015/06/the-internet-of-things-you-dont-really-need/396485.

Bonderud, Douglas. "IoT Security: The Cold, Hard Truth About Digital Device Defense." *Security Intelligence*, November 17, 2017. https://securityintelligence.com/iot-security-the-cold-hard-truth-about-digital-device-defense.

Boston Dynamics, "WildCat." Accessed May 1, 2018. https://www.bostondynamics.com/wildcat.

Boyd, Clark. "What Will the Future of Google Search Results Pages Look Like?" *Search Engine Watch*, April 27, 2017. https://searchenginewatch.com/2017/04/27/what-will-the-future-of-google-search-results-pages-look-like.

Boyle, John. "Calling 911 on a Cell? They Won't Know Your Address." *Citizen Times,* February 23, 2015. http://www.citizen-times.com/story/news/local/2015/02/23/calling-cell-know-address/23893109.

Buntz, Brian. "The World's 5 Smartest Cities," *Internet of Things Institute,* May 18, 2016. http://www.ioti.com/smart-cities/world-s-5-smartest-cities.

Branigan, Tania. "China's Cultural Revolution: son's guilt over the mother he sent to her death." *The Guardian,* March 27, 2013. https://www.theguardian.com/world/2013/mar/27/china-cultural-revolution-sons-guilt-zhang-hongping.

Brown, Evelyn A. "NIST: Performance of Facial Recognition Software Continues to Improve." *NIST* News, June 3, 2014. https://www.nist.gov/news-events/news/2014/06/nist-performance-facial-recognition-software-continues-improve.

Brownell, Kathryn Cramer. "This is How Presidential Campaign Ads First Got on TV." *Time,* August 30, 2016. http://time.com/4471657/political-tv-ads-history.

Brynjolfsson, Erik, and Andrew McAfee. "The Business of Artificial Intelligence." *Harvard Business Review,* July 2017. https://hbr.org/cover-story/2017/07/the-business-of-artificial-intelligence.

Burhouse, Susan, and Yazmin Osaki. "2011 FDIC National Survey of Unbanked and Underbanked Households." Federal Deposit Insurance Corporation, September 2012. https://www.fdic.gov/householdsurvey/2012_unbankedreport.pdf.

Burrows, Leah. "New RoboBee flies, dives, swims and explodes out of the water." *Harvard School of Engineering and Applied Sciences News & Events*, October 25, 2017. https://www.seas.harvard.edu/news/2017/10/new-robobee-flies-dives-swims-and-explodes-out-of-water.

Burns, Matt. "Intel powered the drones during Lady Gaga's Super Bowl halftime show." *TechCrunch*, February 5. 2017. https://techcrunch.com/2017/02/05/intel-powered-the-drones-during-lady-gagas-super-bowl-halftime-show.

Calhoun, Lisa. "3 Reasons Why the Drone Industry Is Hiring Thousands of New Workers," *Inc.,* Sep, 18, 2017. https://www.inc.com/lisa-calhoun/how-big-is-the-drone-job-boom.html.

"Can genes be patented?" *U.S. National Library of Medicine.* Accessed December 13, 2017. https://ghr.nlm.nih.gov/primer/testing/genepatents.

Canary. "Burglars Caught by Canary." *YouTube*, May 16, 2016. https://www.youtube.com/watch?v=ps4xZ4dJZEA.

Cartwright, Mark. "Cohortes Urbanae." *Ancient History Encyclopedia*, December 16, 2016. https://www.ancient.eu/Cohortes_Urbanae.

Castelvecchi, Davide. "Can We Open the Black Box of AI?" *Nature* 538 (October 2016): 20–23. doi: 0.1038/538020a.

Cell Press. "Show Me Your DNA and I'll Tell You Your Eye Color." *ScienceDaily*, March 11, 2009. www.sciencedaily.com/releases/2009/03/090309121933.htm.

Chang, Diana, et al. "A meta-analysis of genome-wide association studies identifies 17 new Parkinson's disease risk loci." *Nature Genetics* (September 2017): 1511-16. doi:10.1038/ng.3955.

Chappell, Bill, and Maggie Penman. "Ransomware Attacks Ravage Computer Networks in Dozens of Countries." *NPR*, May 12, 2017. http://www.npr.org/sections/thetwo-way/2017/05/12/528119808/large-cyber-attack-hits-englands-nhs-hospital-system-ransoms-demanded.

Chavey, S.M. "Be on your best behavior because thousands of security cameras will show Super Bowl activity." *Twin Cities Pioneer Press*, February 1, 2018. https://www.twincities.com/2018/01/30/be-on-your-best-behavior-because-thousands-of-security-cameras-will-show-super-bowl-activity.

Cheng, Jacqui. "Only 0.3% of files on BitTorrent confirmed to be legal." *ArsTechnica*, July 23, 2010. https://arstechnica.com/tech-policy/2010/07/only-03-of-files-on-bit-torrent-confirmed-to-be-legal.

Christakis, Nicholas A. and James H. Fowler. "Social Network Sensors for Early Detection of Contagious Outbreaks." *PLoS* ONE 5.9 (September 2010). doi: 10.1371/journal.pone.0012948.

Ciccatelli, Amanda. "Selling Customer Data in Bankruptcy." *Inside Counsel*, July 9, 2016. http://web3.insidecounsel.com/2016/07/19/selling-customer-data-in-bankruptcy.

Clardy, Jon, et al. "The natural history of antibiotics." *Current Biology* (June 2009): R437-R441. doi:10.1016/j.cub.2009.04.001.

The College of Physicians of Philadelphia. "History of the Immunization Schedule." Accessed December 13, 2017. https://www.historyofvaccines.org/content/history-immunization-schedule.

CNBC. "Alphabet Adds Patent Claim to Uber Intellectual Property Theft Lawsuit." *Reuters*, March 13, 2017. http://www.cnbc.com/2017/03/13/alphabet-adds-patent-claim-to-uber-intellectual-property-theft-lawsuit.html.

CNIL. "Portée du Déréférencement de M. Plaignant Appliqué par Google." March 24, 2016. https://www.cnil.fr/fr/infographie-portee-du-dereferencement-de-mplaignant-applique-par-google.

CNRFID. "Introduction to the RFID." Accessed June 4, 2018. http://www.centrenational-rfid.com/introduction-to-the-rfid-article-15-gb-ruid-202.html.

Coghlan, Andy. "Just 2.5% of DNA Turns Mice into Men." *New Scientist*, May 30, 2002. https://www.newscientist.com/article/dn2352-just-2-5-of-dna-turns-mice-into-men.

______. "Selling the family secrets." *New Scientist,* December 5, 1998. https://www.newscientist.com/article/mg16021633-300-selling-the-family-secrets.

Collins, Keith. "Victims of the WannaCry Ransomware Attacks Have Stopped Paying Up." QZ.com, May 17, 2017. https://qz.com/986094/wannacry-ransomware-attacks-victims-have-stopped-paying-the-ransom.

Columbus, Courtney. "Could Drones Help Save People in Cardiac Arrest?" *NPR,* June 13, 2017. http://www.npr.org/sections/health-shots/2017/06/13/532639836/could-drones-help-save-people-in-cardiac-arrest.

Conniff, Richard. "What the Luddites Really Fought Against." *Smithsonian Magazine*, March 2011. https://www.smithsonianmag.com/history/what-the-luddites-really-fought-against-264412.

Conseil D'État. "Right to Be Delisted." February 24, 2017. http://english.conseil-etat.fr/Activities/Press-releases/Right-to-be-delisted.

Constine, John. "Facebook kills off exact location sharing in Nearby Friends, adds 'Wave.'" *TechCrunch*, December 22, 2016. https://techcrunch.com/2016/12/22/facebook-wave.

______. "Instagram Stories hits 250M daily users, adds Live video replays." *TechCrunch*, June 20, 2017. https://techcrunch.com/2017/06/20/instagram-live-video-replays.

Couzin-Frankel, Jennifer. "Trust Me, I'm a Medical Researcher." *Science* (January 2015): 501–503. doi:10.1126/science.347.6221.501.

Cubrilovic, Nik. "RockYou Hack: From Bad to Worse." *TechCrunch*, December 15, 2009. https://techcrunch.com/2009/12/14/rockyou-hack-security-myspace-facebook-passwords."

Culpan, Tim. "Google Won't Play in China." Bloomberg Gadfly, February 6, 2017. https://www.bloomberg.com/gadfly/articles/2017-02-06/google-china-app-entry-won-t-be-something-to-shout-about.

Darrow, Barb. "Microsoft and Google Give Startups Options to Amazon's Cloud." *Fortune*, August 17, 2017. http://fortune.com/2017/08/17/microsoft-google-amazon-startup-cloud-options.

de Pizan, Christine. *The Book of the Body Politic.* Cambridge: Cambridge University Press, 1994.

de Tocqueville, Alexis. *Democracy in America.* Edited by Harvey C. Mansfield and Delba Winthrop. Chicago: University of Chicago Press, 2000.

De Vynck, Gerrit. "These Search Results Show Why Europe Is Mad at Google." *Bloomberg,* June 27, 2017. https://www.bloomberg.com/news/articles/2017-06-27/these-google-search-results-show-why-europe-is-mad.

DesJardins, Jeff. "Why Hackers Hack: Motives Behind Cyberattacks." *Visual Capitalist*, January 3, 2018. http://www.visualcapitalist.com/hackers-hack-motives-behind-cyberattacks.

Deloitte. "A New Future for R&D: Measuring the return from pharmaceutical innovation 2017." 2018. https://www2.deloitte.com/content/dam/Deloitte/uk/Documents/life-sciences-health-care/deloitte-uk-measuring-roi-pharma.pdf.

Devlin, Hannah. "New Alzheimer's test can predict age when disease will appear." *The Guardian,* March 22, 2017. https://www.theguardian.com/society/2017/mar/22/new-alzheimers-test-can-predict-age-when-disease-will-appear.

Digital Medial Law Project. "Publication of Private Facts." Accessed July 7, 2017. http://www.dmlp.org/legal-guide/publication-private-facts.

DigitalAttackMap. "What Is a DDoS Attack?" Accessed July 3, 2017. http://www.digitalattackmap.com/understanding-ddos.

Downes, Larry. "GDPR and the End of the Internet's Grand Bargain." *Harvard Business Review*, April 9, 2018. https://hbr.org/2018/04/gdpr-and-the-end-of-the-internets-grand-bargain.

Duhigg, Charles. "Bilking the Elderly, With a Corporate Assist." *New York Times*, May 20, 2007. http://www.nytimes.com/2007/05/20/business/20tele.html.

Duke, Alan. "George Zimmerman's Parents Sue Roseanne Barr over Tweets." *CNN*, March 12, 2014. http://www.cnn.com/2014/03/12/showbiz/george-zimmerman-roseanne-barr-lawsuit.

Dunbar, John. "The 'Citizens United' Decision and Why it Matters." Center for Public Integrity, October 18, 2012. https://www.publicintegrity.org/2012/10/18/11527/citizens-united-decision-and-why-it-matters.

Duncan, David Ewing. "On a Mission to Sequence the Genomes of 100,000 People." *New York Times*, June 7, 2010. http://www.nytimes.com/2010/06/08/science/08church.html.

Duo. "Hundreds of Leading Technology Companies Partner with Duo Security to Provide Zero-Trust Security." April 11, 2018. https://duo.com/about/press/releases/hundreds-of-leading-technology-companies-partner-with-duo-security-to-provide-zero-trust-security.

Eddy, Max. "Android Coupons App Leaks Your Personal Information to Everyone." *PC Magazine*, January 6, 2014. http://securitywatch.pcmag.com/mobile-security/319459-android-coupons-app-leaks-your-personal-information-to-everyone.

Eggimann, Patrick, and Aurelia Tamò. "Taming the Beast: Big Data and the Role of Law." *Big Data and Privacy: Making Ends Meet.* (2013): 27–30.

Electronic Privacy Information Center. "EPIC v. FAA: Challenging the FAA's Failure to Establish Drone Privacy Rules." Accessed July 1, 2017. https://epic.org/privacy/litigation/apa/faa/drones.

______. "Identity Theft." Accessed July 7, 2017. https://epic.org/privacy/idtheft.

______. "Spotlight on Surveillance." Accessed July 1, 2017. https://epic.org/privacy/surveillance/spotlight.

eMarketer. "Total US Ad Spending to See Largest Increase Since 2004." July 2, 2014. https://www.emarketer.com/Article/Total-US-Ad-Spending-See-Largest-Increase-Since-2004/1010982.

______. "Most Smartphone Owners Use Location-Based Services." April 22, 2016. https://www.emarketer.com/Article/Most-Smartphone-Owners-Use-Location-Based-Services/1013863.

Enck, William, et al. "TaintDroid: An Information-Flow Tracking System for Realtime Privacy Monitoring on Smartphones." 9th USENIX Symposium on Operating Systems Design and Implementation, 2010. https://www.usenix.org/legacy/event/osdi10/tech/full_papers/Enck.pdf.

European Commision. "E-evidence: cross-border access to electronic evidence." Accessed June 20, 2018. https://ec.europa.eu/info/policies/justice-and-fundamental-rights/criminal-justice/e-evidence-cross-border-access-electronic-evidence_en.

______. "Factsheet on the 'Right to be Forgotten' Ruling (C-131/12)." 2014. http://ec.europa.eu/justice/data-protection/files/factsheets/factsheet_data_protection_en.pdf.

______. "The Intel Antitrust Case." Accessed July 5, 2017. http://ec.europa.eu/competition/sectors/ICT/intel.html.

Ewers, Justin. "Revoking Civil Liberties: Lincoln's Constitutional Dilemma." *US News*, February 10, 2009. https://www.usnews.com/news/history/articles/2009/02/10/revoking-civil-liberties-lincolns-constitutional-dilemma.

Experian. "Targeted Consumer Mailing Lists." Accessed July 6, 2017. http://www.experian.com/small-business/geographic-targeted-mailing-lists.jsp.

Fanelli, Danielle. "How Many Scientists Fabricate and Falsify Research? A Systematic Review and Meta-Analysis of Survey Data?" *PLoS One* (May 2009). doi:10.1371/journal.pone.0005738.

Farivar, Cyrus. "Governor Angrily Tweets about Local TV Station Flying Drone over His House." *ArsTechnica,* August 1, 2017. https://arstechnica.com/tech-policy/2017/08/governor-angrily-tweets-about-local-tv-station-flying-drone-over-his-house.

______. "How Florida Cops Went Door to Door with Fake Cell Device to Find One Man." *ArsTechnica,* June 4, 2014. https://arstechnica.com/tech-policy/2014/06/how-florida-cops-went-door-to-door-with-fake-cell-device-to-find-one-man.

______. "Judge Rules in Favor of 'Drone Slayer,' Dismisses Lawsuit Filed by Pilot." *ArsTechnica,* March 24, 2017. https://arstechnica.com/tech-policy/2017/03/judge-rules-in-favor-of-drone-slayer-dismisses-lawsuit-filed-by-pilot.

Farmer, Steven. "Blockchain Technologies and the EU 'Right to Be Forgotten'—An Insurmountable Tension?" *International Business Times,* September 7, 2016. http://www.ibtimes.co.uk/blockchain-technologies-eu-right-be-forgotten-insurmountable-tension-1580166.

Feathers, Todd. "Middlesex County Prosecutors Building Murder Case against Dzhokhar Tsarnaev in Officer's Slaying." *Boston.com,* April 25, 2013. https://web.archive.org/web/20130426182245/http://boston.com/metrodesk/2013/04/25/middlesex-county-prosecutors-building-murder-case-against-dzhokhar-tsarnaev-officer-slaying/Z5ERGSAAeLeMaDTpsdLIjK/story.html.

Federal Aviation Administration. "FAA Aerospace Forecast: Fiscal Years 2017-2037." Accessed January 12, 2018. https://www.faa.gov/data_research/aviation/aerospace_forecasts/media/FY2017-37_FAA_Aerospace_Forecast.pdf.

Federal Communicactions Commision. "Enhanced 911—Wireless Services." Accessed July 2, 2017. https://www.fcc.gov/general/enhanced-9-1-1-wireless-services.

______. "Restoring Internet Freedom." Accessed June 20, 2018. https://www.fcc.gov/restoring-internet-freedom.

______. "Filings." Accessed July 2, 2017. https://www.fcc.gov/ecfs/search/filings?sort=date_disseminated,DESC&proceedings_name=07-114.

Federal Trade Commision. "Data Brokers: A Call for Transparency and Accountability." May 2014. https://www.ftc.gov/system/files/documents/reports/data-brokers-call-transparency-accountability-report-federal-trade-commission-may-2014/140527databrokerreport.pdf.

______. "Eli Lilly Settles FTC Charges Concerning Security Breach." January 18, 2002. https://www.ftc.gov/news-events/press-releases/2002/01/eli-lilly-settles-ftc-charges-concerning-security-breach.

______. "Facebook Settles FTC Charges That It Deceived Consumers By Failing To Keep Privacy Promises." November 29, 2011. https://www.ftc.gov/news-events/press-releases/2011/11/facebook-settles-ftc-charges-it-deceived-consumers-failing-keep.

______. "Gateway Learning Settles FTC Privacy Charges." July 7, 2004. https://www.ftc.gov/news-events/press-releases/2004/07/gateway-learning-settles-ftc-privacy-charges.

______. "Joint Statement of Acting FTC Chairman Maureen K. Ohlhausen and FCC Chairman Ajit Pai on Protecting Americans' Online Privacy,"

______. "Price Discrimination: Robinson-Patman Violations." Accessed July 5, 2017. https://www.ftc.gov/tips-advice/competition-guidance/guide-antitrust-laws/price-discrimination-robinson-patman.

______. "Protecting Consumer Privacy in an Era of Rapid Change: Recommendations for Businesses and Policymakers." 2012. https://www.ftc.gov/sites/default/files/documents/reports/federal-trade-commission-report-protecting-consumer-privacy-era-rapid-change-recommendations/120326privacyreport.pdf.

Feldman, Ari. "23andMe's Next Phase: Using Customers' DNA to do Drug Research." *Forward,* April 26, 2017. https://forward.com/news/national/383434/23andmes-next-phase-using-customers-dna-to-do-drug-research.

Financial Times. "Unethical Pharma." July 4, 2012. https://www.ft.com/content/94e3da58-c5f1-11e1-a3d5-00144feabdc0.

Firefox. "Lightbeam for Firefox." *YouTube*, October 25, 2013. https://www.youtube.com/watch?v=PvqGy9wz_wA.

Firozi, Paulina. "Internet users raise funds to buy lawmakers' browsing histories in protest." *The Hill*, March 30, 2017. http://thehill.com/policy/technology/326462-internet-users-raise-more-than-200000-to-buy-lawmakers-browsing-histories.

Fisher, Daniel. "Supreme Court Rejects Human-Gene Patents—Sort Of." *Forbes*, June 13, 2013. https://www.forbes.com/sites/danielfisher/2013/06/13/supreme-court-rejects-human-gene-patents-sort-of.

FLIR. "Black Hornet PRS: Airborne Personal Reconnaissance System (PRS)." Accessed June 15, 2018. https://www.flir.com/products/black-hornet-prs.

Fox, Erin. "How Pharma Companies Game the System to Keep Drugs Expensive." *Harvard Business Review,* April 6, 2017. https://hbr.org/2017/04/how-pharma-companies-game-the-system-to-keep-drugs-expensive.

Fox News. "Drone used to find missing woman in North Carolina cornfield, video shows," Nov. 8, 2017. http://www.foxnews.com/tech/2017/11/08/drone-used-to-find-missing-woman-in-north-carolina-cornfield-video-shows.html.

Fox News Tech. "Robot Hummingbird Spy Drone Flies for Eight Minutes, Spies on Bad Guys." Februrary 18, 2011. http://www.foxnews.com/tech/2011/02/18/robot-hummingbird-spy-drone-flies-minutes-spies-bad-guys.html.

Frankel, Alison. "No 'Right to Be Forgotten' Even if Record Is Expunged: 2nd Circuit." *Reuters,* January 28, 2015. http://blogs.reuters.com/alison-frankel/2015/01/28/no-right-to-be-forgotten-even-if-record-is-expunged-2nd-circuit.

Free Software Foundation Europe. "European Commision vs. Microsoft: Chronology of the Case." Accessed July 5, 2017. https://fsfe.org/activities/ms-vs-eu/timeline.en.html.

Freedman, Leonard P., et al. "The Economics of Reproducibility in Preclinical Research." *PLoS Biology* (June 2015). doi:10.1371/journal.pbio.1002165.

Freedom House. "Silencing the Messenger: Communication Apps Under Pressure." November 2016. https://freedomhouse.org/report/freedom-net/freedom-net-2016.

Frier, Sarah. "Facebook Moves to Block Ads on Fake News, Offensive Videos." *Bloomberg,* September 13, 2017. https://www.bloomberg.com/news/articles/2017-09-13/facebook-will-shut-off-ads-for-publishers-that-violate-standards.

Fung, Brian. "Net neutrality takes effect today. Here's how it affects you." *Washington Post,* June 12, 2015. https://www.washingtonpost.com/news/the-switch/wp/2015/06/12/net-neutrality-takes-effect-today-heres-how-it-affects-you.

Future of Privacy Forum. "Smart Cities Resources." Accessed July 3, 2017. https://fpf.org/smart-cities-resources.

Garcia, Mike. "Easy Ways to Build a Better P@$5w0rd." *NIST Blog,* October 4, 2017. https://www.nist.gov/blogs/taking-measure/easy-ways-build-better-p5w0rd.

Garfinkel, Simson. *Database Nation: The Death of Privacy in the 21st Century.* Cambridge: O'Reilly, 2000.

Gartner. "Gartner Forecasts Worldwide Cloud-Based Security Services to Grow 21 Percent in 2017." June 13, 2017. https://www.gartner.com/newsroom/id/3744617.

Geenens, Pascal. "BrickerBot only attacks compromised devices." *Radware Blog,* May 18, 2017. https://blog.radware.com/security/2017/05/brickerbot-attacks-compromised-devices.

Gellman, Robert. "Privacy in the Clouds: Risks to Privacy and Confidentiality from Cloud Computing." *World Privacy Forum,* February 23, 2009. http://www.worldprivacyforum.org/wp-content/uploads/2009/02/WPF_Cloud_Privacy_Report.pdf.

Gemalto. "Data Breach Statistics – Breach Level Index." Accessed June 20, 2018. http://breachlevelindex.com.

Gewirtz, David. "The best password managers for 2018." *CNET*, February 16, 2018. https://www.cnet.com/news/the-best-password-managers-directory.

GfK. "Willingness to share personal data in exchange for benefits or rewards." January 2017. https://www.gfk.com/fileadmin/user_upload/country_one_pager/NL/images/Global-GfK_onderzoek_-_delen_van_persoonlijke_data.pdf.

Ginsberg, Jeremy, et al. "Detecting Influenza Epidemics Using Search Engine Query Data." *Nature* 457 (February 2009): 1012-14. doi: 10.1038/nature07634.

Giuffrida, Joseph P. "Wearables and Application Solutions for Parkinson's Disease: An Overview."MovementDisorders.Org. June 18, 2015. http://www.movementdisorders.org/Congress-2015-Files/pdfs/MDS_2015_TechTaskForce_Giuffrida_E.pdf.

Glaser, April. "If You Go Near the Super Bowl, You Will Be Surveilled Hard." *Wired*, January 31, 2016. https://www.wired.com/2016/01/govs-plan-keep-super-bowl-safe-massive-surveillance.

Goel, Vindu. "Facebook to Let Users Alter Their Ad Profiles." *New York Times*, June 12, 2014. https://www.nytimes.com/2014/06/13/technology/facebook-to-let-users-alter-their-ad-profiles.html.

Goldgeir, Kathy. "Ukraine Cyberattack Wasn't Really About Ransom, Security Experts Say." *NPR*, June 30, 2017. http://www.npr.org/sections/thetwo-way/2017/06/30/535043320/ukraine-cyberattack-wasnt-really-about-ransom-security-experts-say.

Goldmacher, Shane. "Hillary Clinton's 'Invisible Guiding Hand.'" *Politico*, September 7, 2016. http://www.politico.com/magazine/story/2016/09/hillary-clinton-data-campaign-elan-kriegel-214215.

Google. "Developer Guide." Accessed June 20, 2018. https://developers.google.com/maps/documentation/geolocation/intro.

______. "Google Dashboard." Accessed September 20, 2017. https://myaccount.google.com/dashboard.

______. "Google Transparency Report." Accessed September 9, 2017. https://transparencyreport.google.com.

______. "How Search Algorithms Work." Accessed September 9, 2017. https://www.google.com/search/howsearchworks/algorithms.

______. "Latitude has been retired." Accessed August 14, 2015. https://web.archive.org/web/20150814192105/https://support.google.com/gmm/answer/3001634?p=maps_android_latitude&rd=1.

______. "Multi-regional and multilingual sites." Accessed December 13, 2017. https://support.google.com/webmasters/answer/182192.

______. "My Activity." Accessed September 9, 2017. https://myactivity.google.com/myactivity.

______. "Reconsideration Requests." Accessed September 9, 2017. https://support.google.com/webmasters/answer/35843.

______. "Search Removals Under European Privacy Law." Accessed September 9, 2017. https://www.google.com/transparencyreport/removals/europeprivacy.

______. "Webmaster Guidelines." Accessed September 9, 2017. https://support.google.com/webmasters/answer/35769.

Google Inc. v. CNIL (C-507/17). http://eur-lex.europa.eu/legal-content/EN/TXT/?uri=CELEX:62017CN0507.

Graff, David. "An Update to Our AdWords Policy on Lending Products." Keyword. May 11, 2016. https://www.blog.google/topics/public-policy/an-update-to-our-adwords-policy-on.

Greenberg, Andy. "How Secure is the IPhone X's FaceID? Here's What We Know." *Wired,* September 12, 2017. https://www.wired.com/story/iphone-x-faceid-security.

______. "The Saboteurs of Search." *Forbes*, June 28, 2007. https://www.forbes.com/2007/06/28/negative-search-google-tech-ebiz-cx_ag_0628seo.html.

Greenstone, Gerry. "The history of bloodletting." *BC Medical Journal* 52.1 (February 2010): 12-14. http://www.bcmj.org/premise/history-bloodletting.

Gregory, Derek. "How a Hollywood Film Reveals the Reality of Drone Warfare." *The Guardian,* April 9, 2016. https://www.theguardian.com/world/2016/apr/10/drone-warfare-eye-in-the-sky.

GridLocate. "GridLocate." Accessed July 2, 2017. https://gridlocate.com.

Groden, Claire. "What Is the 'Right to Be Forgotten' and Will It Ever Exist in the United States?" *Inverse*, May 19, 2016. https://www.inverse.com/article/15924-what-is-the-right-to-be-forgotten-and-will-it-ever-exist-in-the-united-states.

Guliani, Neema Singh, and Naureen Shah. "The CLOUD Act Doesn't Help Privacy and Human Rights: It Hurts Them." *Lawfare*, March 16, 2018. https://www.lawfareblog.com/cloud-act-doesnt-help-privacy-and-human-rights-it-hurts-them.

Gymrek, Melissa, et al. "Identifying Personal Genomes by Surname Inference." *Science* (January 2013): 321–324. doi:10.1126/science.1229566.

"H.R. 1313: Preserving Employee Wellness Programs Act." Accessed December 13, 2017. https://www.congress.gov/bill/115th-congress/house-bill/1313.

H., Victor. "The phones before the iPhone: 10 iconic devices." *PhoneArena.com*, October 26, 2014. https://www.phonearena.com/news/The-phones-before-the-iPhone-10-iconic-devices_id62159.

Harris, Richard. "Drugs That Work in Mice Often Fail When Tried in People." *NPR*, April 10, 2017. https://www.npr.org/sections/health-shots/2017/04/10/522775456/drugs-that-work-in-mice-often-fail-when-tried-in-people.

Hatmaker, Taylor. "Consolidated Appropriations Act of 2018." *TechCrunch*, March 22, 2018. https://techcrunch.com/2018/03/22/cloud-act-omnibus-bill-house.

Hautala, Laura. "Internet-Connected Homes Open the Door to Hackers." *CNET,* December 28, 2015. https://www.cnet.com/news/internet-connected-homes-open-the-door-to-hackers.

———. "Smart Toy Flaws Make Hacking Kids' Info Child's Play." *CNET,* February 28, 2017. https://www.cnet.com/news/cloudpets-iot-smart-toy-flaws-hacking-kids-info-children-cybersecurity.

———. "Why It Was So Easy to Hack the Cameras that Took Down the Web." *CNET,* October 24, 2016. Accessed July 3, 2017. https://www.cnet.com/how-to/ddos-iot-connected-devices-easily-hacked-internet-outage-webcam-dvr.

Hattem, Julian. "FBI Director: Cover Up Your Webcam." *The Hill,* September 14, 2016. http://thehill.com/policy/national-security/295933-fbi-director-cover-up-your-webcam.

Hauser, Christine. "Google to Ban All Payday Loan Ads." *New York Times*, May 11, 2016. https://www.nytimes.com/2016/05/12/business/google-to-ban-all-payday-loan-ads.html.

Hayes, Heather B. "How the Cloud Improves Transparency at the State Department, USAID and NOAA." *FedTech*, July 25, 2017. https://fedtechmagazine.com/article/2017/07/how-cloud-improves-transparency-state-department-usaid-and-noaa.

Hecht, Jeff. "FCC Sets Rules for Copper Phase Out." *IEEE Spectrum*, August 7, 2015. http://spectrum.ieee.org/tech-talk/telecom/internet/fcc-sets-rules-for-copper-phase-out.

HelloSpy. "HelloSpy." Accessed July 2, 2017. http://hellospy.com/homepage.aspx.

Herper, Matthew. "The Truly Staggering Cost of Inventing New Drugs." *Forbes*, February 10, 2012. https://www.forbes.com/sites/matthewherper/2012/02/10/the-truly-staggering-cost-of-inventing-new-drugs.

Hill, Kashmir. "How Target Figured Out a Teen Girl Was Pregnant Before Her Father Did." *Forbes*, February 16, 2012. https://www.forbes.com/sites/kashmirhill/2012/02/16/how-target-figured-out-a-teen-girl-was-pregnant-before-her-father-did/2/#69eba4ac47e8.

Hirschler, Ben. "Amgen buys Icelandic gene hunter Decode for $415 million." *Reuters,* December 10, 2012. https://www.reuters.com/article/us-amgen-decode/amgen-buys-icelandic-gene-hunter-decode-for-415-million-idUSBRE8B90IU20121210.

History.com. "William Randolph Hearst Stops *Citizen Kane* Ads." January 8, 2009. http://www.history.com/this-day-in-history/william-randolph-hearst-stops-citizen-kane-ads.

Hood, James R. "FAA Gives Commercial Drones Takeoff Clearance." *Consumer Affairs*, June 21, 2016. https://www.consumeraffairs.com/news/faa-gives-commercial-drones-takeoff-clearance-062116.html.

Horowitz, Michael. "Google knows nearly every Wi-Fi password in the world." *Computerworld*, September 12, 2013. https://www.computerworld.com/article/2474851/android/android-google-knows-nearly-every-wi-fi-password-in-the-world.html.

Horrigan, John B. "Use of Cloud Computing Applications and Services." *Pew Internet*, September 12, 2008. http://www.pewinternet.org/2008/09/12/use-of-cloud-computing-applications-and-services.

Hughes, Laura. "Google bans adverts alongside 'extremely offensive' YouTube videos." *Telegraph*, June 19, 2017. http://www.telegraph.co.uk/news/2017/06/18/google-bans-adverts-alongside-extremely-offensive-youtube-videos.

Hulu. "AWS Chosen by Hulu as Its Cloud Provider." *Business Wire,* August 14, 2017. https://www.businesswire.com/news/home/20170814005141/en.

Humphries, Daniel. "US Attitudes Toward the 'Right to Be Forgotten.'" *Software Advice*, September 5, 2014. http://www.softwareadvice.com/security/industryview/right-to-be-forgotten-2014.

Hurley, Lawrence. "US Supreme Court to Settle Major Cellphone Privacy Case." *Reuters,* June 5, 2017. http://www.reuters.com/article/us-usa-court-mobilephone-idUSKBN18W1RY.

Hussain, Kumail. "Cloud storage now more affordable: Announcing general availability of Azure Archive Storage." *Microsoft Azure Blog*, December 13, 2017. https://azure.microsoft.com/en-us/blog/cloud-storage-now-more-affordable-announcing-general-availability-of-azure-archive-storage.

Hvistendahl, Mara. "China's Theft of US Trade Secrets under Scrutiny." *Science,* February 28, 2017. doi:10.1126/science.aal0837.

ICO. "Types of encryption." Accessed June 4, 2018. https://ico.org.uk/for-organisations/guide-to-data-protection/encryption/types-of-encryption.

International Commissioner's Office. "The Right to Erasure (The Right to Be Forgotten)." Accessed June 30, 2017. https://ico.org.uk/for-organisations/data-protection-reform/overview-of-the-gdpr/individuals-rights/the-right-to-erasure.

Internet Live Stats. "Google Search Statistics." Accessed July 5, 2017. http://www.internetlivestats.com/google-search-statistics.

"Interstate Commerce Act (1887)." *Our Documents.* Accessed June 4, 2018. https://www.ourdocuments.gov/doc.php?flash=true&doc=49.

Ioannidis, John P. A. "Contradicted and Initially Stronger Effects in Highly Cited Clinical Research." *The Journal of the American Medical Association* 294 (2005): 218-28. doi:10.1001/jama.294.2.218.

______. "Why Most Published Research Findings Are False." *PLoS Medicine* (August 2005): 0696-0701. doi: 10.1371/journal.pmed.0020124.

IPCommision.org. "Update to the IP Commission Report 2017." The National Bureau of Asian Research. February 2017. http://www.ipcommission.org/report/IP_Commission_Report_Update_2017.pdf.

Jablonski, Chris. "Getting a Fix on Indoor GPS." *ZDNet,* October 7, 2011. http://www.zdnet.com/article/getting-a-fix-on-indoor-gps.

Jackson, Nicholas. "10 Notorious Google Bombs." *The Atlantic,* February 25, 2011. https://www.theatlantic.com/technology/archive/2011/02/10-notorious-google-bombs/71731.

Jacobson, Roni. "Many Antidepressant Studies Found Tainted by Pharma Company Influence." *Scientific America*, October 21, 2015. https://www.scientificamerican.com/article/many-antidepressant-studies-found-tainted-by-pharma-company-influence.

Jain, Aniketh. "Cloud communication helping startups grow, creating a win-win scenario." *YourStory/MyStory*, September 2017. https://yourstory.com/mystory/42ecabce96-cloud-communication-he.

Jansen, Bart. "NYC firefighters use drone to help battle blaze for first time." *USA Today,* March 8, 2017. https://www.usatoday.com/story/news/2017/03/08/drone-firefighters/98848038.

Johns Hopkins Medicine. "The Legacy of Henrietta Lacks." Accessed December 13, 2017. https://www.hopkinsmedicine.org/henriettalacks/index.html.

Jones, Meg Leta. *Ctrl + Z The Right to Be Forgotten.* New York: NYU Press, 2016.

Kahn, David. *The Codebreakers: The Comprehensive History of Secret Communication from Ancient Times to the Internet.* New York: Scribner, 1996.

Kant, Immanuel. *Grounding for the Metaphysics of Morals.* Translated by James W. Ellington. Indianapolis: Hackett Publishing Company, 1981.

Karr, Timothy. "Net Neutrality Violations: A Brief History." *Free Press,* January 24, 2018. https://www.freepress.net/our-response/expert-analysis/explainers/net-neutrality-violations-brief-history.

Kastrenakes, Jacob. "YouTube will hide, but not remove, some 'supremacist' videos." *The Verge,* August 1, 2017. https://www.theverge.com/2017/8/1/16072840/youtube-placing-limits-on-supremacist-videos.

Kats, Rimma. "The End to Data as We Know It?" eMarketer, 2017. https://www.emarketer.com/Article/End-Data-Know-It/1015018.

Kaufman, Stacy, et al. "Micropayments: A Viable Business Model?" *Stanford University CS 181,* Spring 2011. http://cs.stanford.edu/people/eroberts/cs201/projects/2010-11/MicropaymentsAndTheNet/index.html.

Kelland, Kate. "Big Pharma donates drugs for neglected diseases." *Reuters,* January 30, 2012. https://www.reuters.com/article/us-diseases-neglected-pharma/big-pharma-donates-drugs-for-neglected-diseases-idUSTRE80T0LO20120130.

Keller, Daphne. "Global Right to Be Forgotten Delisting: Why CNIL Is Wrong." *Stanford Law School: The Center for Internet and Society,* November 18, 2016. http://cyberlaw.stanford.edu/blog/2016/11/global-right-be-forgotten-delisting-why-cnil-wrong.

Keller, Jared. "How The CIA Uses Social Media to Track How People Feel." *The Atlantic,* November 4, 2011. https://www.theatlantic.com/technology/archive/2011/11/how-the-cia-uses-social-media-to-track-how-people-feel/247923.

Kelly, Erin. "Senate Approves USA Freedom Act." *Standard-Times,* June 2, 2015. https://www.gosanangelo.com/story/news/politics/2015/06/02/patriot-act-usa-freedom-act-senate-vote/28345747.

Kelly, John, and Brendan Keefe. "911's Deadly Flaw: Lack of Location Data." *USA Today,* February 22, 2015. https://www.usatoday.com/story/news/2015/02/22/cellphone-911-lack-location-data/23570499.

Kennedy, John. "$1.4bn Investment in Blockchain Start-Ups in Last 9 Months, Says PwC Expert." *Silicon Republic,* November 4, 2016. https://www.siliconrepublic.com/start-ups/blockchain-pwc-investment.

Kerr, Orin. "One law professor's overview of the confusing net neutrality debate." *Washington Post,* November 28, 2017. https://www.washingtonpost.com/news/volokh-conspiracy/wp/2017/11/28/one-law-professors-overview-of-the-confusing-net-neutrality-debate.

Key, Craig B., et al. "Can First Responders Be Sent to Selected 9-1-1 Emergency Medical Services Calls without an Ambulance?" *Academic Emergency Medicine* 10 (April 2003): 339-46. doi:10.1111/j.1553-2712.2003.tb01346.x.

Kiger, Patrick J. "How They Identified the Bombers: A Timeline From Event to Capture," National Geographic, April 1, 2014. http://web.archive.org/web/20171112235546/http://channel.nationalgeographic.com/inside-the-hunt-for-the-boston-bombers/articles/how-they-identified-the-bombers.

Knight, Shawn. "Equifax reveals full extent of last year's massive data breach." *TechSpot,* May 8, 2018. https://www.techspot.com/news/74523-equifax-reveals-full-extent-last-year-massive-data.html.

KNIME. "KNIME." Accessed August 4, 2017. https://www.knime.org.

Konkel, Frank. "CIA Can Anticipate Social Unrest 'Three to Five Days' Out in Some Cases." *NextGov,* October 4, 2016. http://www.nextgov.com/defense/2016/10/cia-can-predict-social-unrest-three-five-days-out/132102.

Kotsiantis, S. B., Zaharakis, I.D., and P. E. Pintelas. "Machine Learning: A Review of Classification." *Artificial Intelligence Review* 26 (2006). doi: 10.1007/s10462-007-9052-3.

Kottasova, Ivana. "Europe Takes on Apple, Facebook, Google and Amazon." *CNN Money,* April 3, 2015. http://money.cnn.com/2015/04/03/news/companies/europe-google-apple-facebook-amazon/index.html.

Koziol, Michael. "It's Never Too Early To Think About 6G." *IEEE Spectrum*, May 22, 2018. https://spectrum.ieee.org/tech-talk/telecom/wireless/its-never-too-early-to-think-about-6g.

Kram, Zach. "The Postseason of Our 3-Point Discontent." *The Ringer*, May 31, 2018. https://www.theringer.com/nba/2018/5/31/17411902/nba-playoffs-3-point-success.

Krebs, Brian. "Equifax Breach: Setting the Record Straight." *Krebs on Security*, September 20, 2017. https://krebsonsecurity.com/2017/09/equifax-breach-setting-the-record-straight.

Kreps, Sarah E. *Drones: What Everyone Needs to Know*, New York: Oxford University Press, 2016.

Krucoff, Carol. "The 6 O'Clock Scholar." *New York Times*, January 29, 1984. https://www.washingtonpost.com/archive/lifestyle/1984/01/29/the-6-oclock-scholar/eed58de4-2dcb-47d2-8947-b0817a18d8fe.

Kundra, Vivek. "Federal Cloud Computing Strategy." *DHS.gov,* February 8, 2011. https://www.dhs.gov/sites/default/files/publications/digital-strategy/federal-cloud-computing-strategy.pdf.

Kutner, Max. "London Firefighters Used Drone to Battle Grenfell Tower Blaze." *Newsweek*, June 6, 2017. http://www.newsweek.com/london-grenfell-tower-fire-departments-drones-626824.

Lafky, Deborah. "The Safe Harbor Method of De-Identification: An Empirical Test." Global Health Care, LLC, October 8, 2009. http://www.ehcca.com/presentations/HIPAAWest4/lafky_2.pdf.

Lafrance, Adrienne. "When People Feared Computers." *The Atlantic*, March 30, 2015. https://www.theatlantic.com/technology/archive/2015/03/when-people-feared-computers/388919.

Lallanilla, Mark. "Are You Carrying the Red Hair Gene?" *Science Daily,* January 28, 2013. https://www.livescience.com/26633-redhead-dna-gene.html.

LaMotte, Sandee. "Scott Gottlieb: Conflicts surround Trump's FDA pick." *CNN,* April 4, 2017. http://www.cnn.com/2017/04/04/health/fda-gottlieb-background-qualifications.

Landers, Jackson. "In the Early 19th Century, Firefighters Fought Fires … and Each Other." *Smithsonian Magazine*, September 27, 2016. https://www.smithsonianmag.com/smithsonian-institution/early-19-century-firefighters-fought-fires-each-other.

Lane, Frederick S. *American Privacy: The 400-Year History of Our Most Contested Right.* Boston: Beacon, 2011.

Langreth, Robert. "California Orders Stop to Gene Testing." *Forbes*, June 14, 2008. https://www.forbes.com/2008/06/14/stop-gene-testing-biz-healthcare-cz_rl_0614genetest.html.

Langreth, Robert and Matthew Herper. "States Crack Down on Online Gene Tests." *Forbes,* April 18, 2008. https://www.forbes.com/2008/04/17/genes-regulation-testing-biz-cx_mh_bl_0418genes.html.

Langston, Jennifer. "How well do facial recognition algorithms cope with a million strangers?" *UW News*, June 23, 2016. https://www.washington.edu/news/2016/06/23/how-well-do-facial-recognition-algorithms-cope-with-a-million-strangers.

Lapowsky, Issie. "Trump's Big Data Mind Explains How He Knew Trump Could Win." *Wired,* November 9, 2016. https://www.wired.com/2016/11/trump-polling-data.

Lardinois, Frederic. "Google's Cloud Platform gets a cheaper, lower-performance networking tier." *TechCrunch*, August 23, 2017. https://techcrunch.com/2017/08/23/googles-cloud-platform-gets-a-cheaper-lower-performance-networking-tier.

Larson, Selena. "The hacks that left us exposed in 2017." *CNN Money*, December 20, 2017. https://money.cnn.com/2017/12/18/technology/biggest-cyberattacks-of-the-year/index.html.

Law, Jacky. *Big Pharma: Exposing the Global Healthcare Agenda.* New York: Carroll & Graf Publishers, 2006.

Lawson, Stephen. "Ten Ways Your Smartphone Knows Where You Are." *PC World*, April 6, 2012. https://www.pcworld.com/article/253354/ten_ways_your_smartphone_knows_where_you_are.html.

Lazer, David, and Ryan Kennedy. "What We Can Learn from the Epic Failure of Google Flu Trends." *Wired*, October 1, 2015. https://www.wired.com/2015/10/can-learn-epic-failure-google-flu-trends.

LeddarTech Inc. "Smart Cities – Using Solid State LiDAR Sensors for Traffic Management." *AZO Sensors*, March 28, 2017. https://www.azosensors.com/article.aspx?ArticleID=773.

Levy, Ari. "Google's 'Phantom' Algorithm Update Hits Websites." *NBC News*, May 13, 2015. http://www.nbcnews.com/tech/internet/googles-phantom-algorithm-update-hits-websites-n358346.

Lewis, Dave. "The DDoS Attack Against Dyn One Year Later." *Forbes*, October 23, 2017. https://www.forbes.com/sites/davelewis/2017/10/23/the-ddos-attack-against-dyn-one-year-later.

Lewis, Paul and Spencer Ackerman. "NSA: Dianne Feinstein Breaks Ranks to Oppose US Spying on Allies." *The Guardian*, October 29, 2013. https://www.theguardian.com/world/2013/oct/28/nsa-surveillance-dianne-feinstein-opposed-allies.

Lewis, Paul. "You're being watched: there's one CCTV camera for every 32 people in UK." *The Guardian*, March 2, 2011. https://www.theguardian.com/uk/2011/mar/02/cctv-cameras-watching-surveillance.

Lewis, Tanya. "Human Genome Marks 10th Anniversary." *Live Science*, April 14, 2013. https://www.livescience.com/28708-human-genome-project-anniversary.html.

Leyser, Ottoline. "The dark side of research – when chasing prestige becomes the prize." *The Conversation*, December 4, 2014. https://theconversation.com/the-dark-side-of-research-when-chasing-prestige-becomes-the-prize-35001.

Lin, Patrick. "Blockchain: The Missing Link Between Genomics and Privacy?" *Forbes*, May 8, 2017. https://www.forbes.com/sites/patricklin/2017/05/08/blockchain-the-missing-link-between-genomics-and-privacy/#2e926d524b77.

Lippert, Christopher, et al. "Identification of individuals by trait prediction using whole-genome sequencing data." *PNAS* (September 2017): 10166-10171. doi: 10.1073/pnas.1711125114.

Lomas, Natasha. "Dyn DNS DDoS Likely the Work of Script Kiddies, Says Flashpoint." *TechCrunch*, October 26, 2016. https://techcrunch.com/2016/10/26/dyn-dns-ddos-likely-the-work-of-script-kiddies-says-flashpoint.

______. "French Senate Backs Bid to Force Google to Disclose Search Algorithm Workings." *TechCrunch,* April 17, 2017. https://techcrunch.com/2015/04/17/french-senate-backs-bid-to-force-google-to-disclose-search-algorithm-workings/.

Lumb, Richard. "Downside of Bitcoin: A Ledger That Can't Be Corrected." *New York Times*, September 9, 2016. https://www.nytimes.com/2016/09/10/business/dealbook/downside-of-virtual-currencies-a-ledger-that-cant-be-corrected.html.

Luna, Taryn. "FBI Sifting for Clues in Mountains of Evidence." *Boston Globe*, April 17, 2013. https://www.bostonglobe.com/business/2013/04/16/software-tools-offer-limited-help-fbi-agents-analyzing-marathon-video/mXM8fYyBDY8B05iJj1UaLN/story.html.

MacAskill, Ewen. "Yahoo Files Lawsuit against NSA Over User Data Requests." *The Guardian,* September 9, 2013. https://www.theguardian.com/world/2013/sep/09/yahoo-lawsuit-nsa-surveillance-requests.

Manjoo, Farhad. "Facebook Followed You to the Supermarket." *Slate,* March 20, 2013. http://www.slate.com/articles/technology/technology/2013/03/facebook_advertisement_studies_their_ads_are_more_like_tv_ads_than_google.single.html.

Marks, Paul. "How Police Catch Drone-Flying Criminals." *BBC,* July 31, 2017. http://www.bbc.com/future/story/20170731-how-cops-catch-drone-flying-criminals.

Marr, Bernard. "How Big Data and Analytics Are Changing Football." *Smart Data Collective,* July 28, 2015. https://www.smartdatacollective.com/how-big-data-and-analytics-are-changing-footbal/.

Mayer-Schönberger, Viktor, and Kenneth Cukier. *Big Data: A Revolution that Will Transform How We Live, Work, and Think.* Boston: Mariner Books, 2014.

McAfee. "Oversharing on Social Media Common Amongst 50+." October 23, 2013. https://securingtomorrow.mcafee.com/consumer/family-safety/50plus-tech-savvy-but-still-at-risk.

McCauley, Lauren. "Thanks to US Congress, Endless War Will…Not End." *Common Dreams,* May 18, 2016. https://www.commondreams.org/news/2016/05/18/thanks-us-congress-endless-war-will-not-end.

McDonald, Aleecia M., and Lorrie Faith Cranor. "The Cost of Reading Privacy Policies." *I/S: A Journal of Law and Policy for the Information Society* 4.3 (Winter 2008): 540-65. http://moritzlaw.osu.edu/students/groups/is/files/2012/02/Cranor_Formatted_Final.pdf.

McFarland, Kevin. "*Eye in the Sky* Is the Quintessential Modern War Film." *Wired,* April 1, 2016. https://www.wired.com/2016/04/eye-in-the-sky-modern-war-film.

McGoogan, Cara, and Danielle Demetriou. "Peace Sign Selfies Could Let Hackers Copy Your Fingerprints." *Telegraph,* January 12, 2017. http://www.telegraph.co.uk/technology/2017/01/12/peace-sign-selfies-could-let-hackers-copy-fingerprints.

McKie, Robin. "Why do identical twins end up having such different lives?" *Guardian,* June 1, 2013. https://www.theguardian.com/science/2013/jun/02/twins-identical-genes-different-health-study.

McMenamin, Dave. "Cavaliers make 25 3-pointers to set NBA regular-season record." *ESPN*, March 4, 2017. http://www.espn.com/nba/story/_/id/18819129/cleveland-cavaliers-make-25-3-pointers-set-nba-regular-season-game-record.

McMillan, Robert. "The Man Who Wrote Those Password Rules Has a New Tip: N3v$r M1^d!" *Wall Street Journal,* August 7, 2017. https://www.wsj.com/articles/the-man-who-wrote-those-password-rules-has-a-new-tip-n3v-r-m1-d-1502124118.

Medical Protection. "Content to consent?" November 1, 2003. https://www.medicalprotection.org/uk/resources/case-reports/case-reports/uk-content-to-consent.

Mell, Peter, and Timothy Grance. "The NIST Definition of Cloud Computing." NIST.Gov. September 2011. http://nvlpubs.nist.gov/nistpubs/Legacy/SP/nistspecialpublication800-145.pdf.

Mepham, Steve. "The Deadzone: Why Big Data Is Dead." *MyCustomer.com*, February 3, 2016. https://www.mycustomer.com/marketing/data/the-deadzone-why-big-data-is-dead.

Mezzofiore, Gianluca. "How a Little-Known Data Firm Helped Trump Become President." *Mashable,* November 10, 2016. http://mashable.com/2016/11/10/donald-trump-polling-data.

Miessler, Daniel. "A Shodan Tutorial and Primer." Updated October 26, 2017. https://danielmiessler.com/study/shodan.

Mikulak, Anna. "Attention on the Brain." Association for Psychological Science, December 30, 2015. http://www.psychologicalscience.org/observer/attention-on-the-brain.

Miller, Claire Cain. "When Algorithms Discriminate." *New York Times*, July 9, 2015. https://www.nytimes.com/2015/07/10/upshot/when-algorithms-discriminate.html.

Miller, Megan. "What the Heck Is Ricin?" *Popular Science,* February 29, 2008. http://www.popsci.com/scitech/article/2008-02/what-heck-ricin.

Miniter, Frank. "Google Sadly Joins the Anti-Gun Brigade." *Forbes*, July 5, 2012. https://www.forbes.com/sites/frankminiter/2012/07/05/google-sadly-joins-the-anti-gun-brigade.

Mirani, Leo. "Does Europe Hate Google or Hate America?" *Quartz*, May 4, 2015. https://qz.com/394273/does-europe-hate-google-or-america.

Mitnick, Kevin D. and William L Simon. *The Art of Intrusion: The Real Stories behind the Exploits of Hackers, Intruders & Deceivers*. Indianapolis: Wiley Publishing, 2005.

Montjoye, Yves-Alexandre de, César A. Hidalgo, Michel Verleysen, and Vincent D. Blondel. "Unique in the Crowd: The Privacy Bounds of Human Mobility." *Nature* (2013). doi:10.1038/srep01376.

Murison, Malek. "Drones are Helping Scientists Weigh Seals in Antarctica." *DroneLife,* November 30, 2017. https://dronelife.com/2017/11/30/drones-scientists-weigh-seals-antarctica.

Nakashima, Ellen, and Joby Warrick. "Stuxnet Was Work of US and Israeli Experts, Officials Say." *Washington Post,* June 9, 2012. https://www.washingtonpost.com/world/national-security/stuxnet-was-work-of-us-and-israeli-experts-officials-say/2012/06/01/gJQAlnEy6U_story.html.

Nasir, Sarwat. "Firefighting drone to put out fires in Dubai in the future." *Khaleej Times*, October 10, 2017. https://www.khaleejtimes.com/news/firefighting-drones-to-put-out-fires-in-dubai-in-the-future.

National Human Genome Research Institute. "Learning About Huntington's Disease." Updated November 17, 2011. https://www.genome.gov/10001215/learning-about-huntingtons-disease.

______. "Privacy in Genomics." April 21, 2015. https://www.genome.gov/27561246/privacy-in-genomics.

National Institute of Neurological Disorders and Stroke. "Gene Therapy for Tay-Sachs Disease." Accessed June 20, 2018. https://www.ninds.nih.gov/Disorders/Clinical-Trials/Gene-Therapy-Tay-Sachs-Disease.

Nature. "Beyond the Genome." *Nature News* (February 2015): 273. doi: 10.1038/518273a.

NBA. "Once again, Houston Rockets set single-season record for most 3-pointers made." *NBA.com,* April 1, 2018. http://www.nba.com/article/2018/04/01/houston-rockets-break-single-season-record-most-3-pointers-made.

NCSL. "Current Unmanned Aircraft State Law Landscape." July 25, 2017. http://www.ncsl.org/research/transportation/current-unmanned-aircraft-state-law-landscape.aspx.

NEC. "Ensuring Digital Safety with Robust Cyber Security." 2014. http://sg.nec.com/en_SG/pdf/brochures/PublicSafety/NEC_Cyber_Security_whitepaper_A4_4pp-FA_OABW.pdf.

Negroponte, Nicholas, and Richard Pawson. "Pay Whom Per What When, Part 1." *Wired,* February 1, 1997. https://www.wired.com/1997/02/negroponte-41.

Nelson, Paul. "Industrywide Standards for Document-Level Security in Enterprise Search." *Search Technologies*. Accessed July 5, 2017. https://www.searchtechnologies.com/blog/document-level-security-standards.

NENA. "9-1-1 Origin and History." Accessed July 2, 2017. http://www.nena.org/?page=911overviewfacts.

NIST. "Cybersecurity Framework." Accessed September 9, 2017. https://www.nist.gov/cyberframework.

______. "Digital Identity Guidelines: Now Available." June 22, 2017. https://pages.nist.gov/800-63-3.

______. "History." Accessed August 4, 2017. https://www.nist.gov/timeline.

______. "Special Publication 800-63 Version 1.0.1." September 2004. https://nvlpubs.nist.gov/nistpubs/Legacy/SP/nistspecialpublication 800-63ver1.0.1.pdf.

Nix, Elizabeth. "Tuskegee Experiment: The Infamous Syphilis Study." *History.com,* May 6, 2017. http://www.history.com/news/the-infamous-40-year-tuskegee-study.

Norse Corp. "Norse." Accessed July 3, 2017. http://www.norse-corp.com.

Nunez, Michael. "What Is 5G and How Will It Make My Life Better?" *Gizmodo*, February 24, 2016. https://gizmodo.com/what-is-5g-and-how-will-it-make-my-life-better-1760847799.

O'Neil, Cathy. *Weapons of Math Destruction: How Big Data Increases Inequality and Threatens Democracy.* New York: Crown, 2016.

OECD. "Thirty Years After the OECD Privacy Guidelines." 2011. http://www.oecd.org/sti/ieconomy/49710223.pdf.

Ohm, Paul. "Broken Promises of Privacy: Responding to the Surprising Failure of Anonymization." *UCLA Law Review* 57 (2010): 1701-77. http://www.uclalawreview.org/pdf/57-6-3.pdf.

Olen, Helaine. "Congratulations! You Lost." *Slate,* May 24, 2016. http://www.slate.com/articles/business/the_bills/2016/05/robocall s_have_triumphed_over_the_do_not_call_list_whose_fault_is_it.ht ml.

Olenski, Steve. "For Consumers, Data is a Matter of Trust." *Forbes,* April 18, 2016. https://www.forbes.com/sites/steveolenski/2016/04/18/for-consumers-data-is-a-matter-of-trust.

Olivarez-Giles, Nathan. "'Pokémon Go' Creator Closes Privacy Hole but Still Collects User Data." *Wall Street Journal,* July 13, 2016. https://www.wsj.com/articles/pokemon-go-creator-closes-privacy-hole-but-still-collects-user-data-1468363704.

Online Etymology Dictionary. "Discrimination." Accessed July 7, 2017. http://www.etymonline.com/index.php?term=discrimination.

OpenSecrets.org. "Pharmaceuticals/Health Products: Industry Profile: Summary, 2017." Accessed December 13, 2017. https://www.opensecrets.org/lobby/indusclient.php?id=h04.

Packard, Vance. "Don't Tell It to the Computer." *New York Times*, January 8, 1967. https://timesmachine.nytimes.com/timesmachine/1967/01/08/835 69846.html?pageNumber=235.

Pasquale, Frank. *The Black Box Society: The Secret Algorithms that Control Money and Information.* Cambridge: Harvard University Press, 2016.

______. "Internet Nondiscrimination Principles: Commercial Ethics for Carriers and Search Engines." *University of Chicago Legal Forum* 1.6 (2008): 263-99. http://chicagounbound.uchicago.edu/cgi/viewcontent.cgi?article=1 427&context=uclf.

Patterson, Dan. "Zero-day exploits: A cheat sheet for professionals." *TechRepublic,* May 8, 2017. https://www.techrepublic.com/article/zero-day-exploits-the-smart-persons-guide.

Paton, Callum. "What is Sarin Gas, The Chemical Weapon that has Killed Over 70 In Syria?" *Newsweek,* April 5, 2017. http://www.newsweek.com/what-sarin-gas-chemical-weapon-has-killed-over-70-syria-579523.

Paul, Deanna. "How Smart Devices Could Violate Your Privacy." *Rolling Stone,* July 24, 2017. http://www.rollingstone.com/culture/features/how-smart-devices-could-violate-your-privacy-w492823.

Paul, Ian. "How to quickly check that your home IoT devices are secure." *PC World*, January 12, 2017. https://www.pcworld.com/article/3155466/internet-of-things/how-to-quickly-check-that-your-home-iot-devices-are-secure.html.

Payton, Theresa, and Ted Claypoole. *Privacy in the Age of Big Data: Recognizing Threats, Defending Your Rights, and Protecting Your Family*. Lanham: Rowman & Littlefield Publishers, 2014.

Pennisi, Elizabeth. "The Genetics of Blond Hair." *Science,* June 1, 2014. http://www.sciencemag.org/news/2014/06/genetics-blond-hair.

Perez, Sarah. "WhatsApp Tests Real-Time Location Sharing in Its App." *TechCrunch,* February 3, 2017. https://techcrunch.com/2017/02/03/whatsapp-tests-real-time-location-sharing-in-its-app.

______. "Plex changes its new privacy policy after backlash, clarified it's not trying to see what's in your library." *TechCrunch*, August 21, 2017. https://techcrunch.com/2017/08/21/plex-changes-its-new-privacy-policy-after-backlash-clarified-its-not-trying-to-see-whats-in-your-library.

The Personal Genome Project. "The Personal Genome Project." Accessed December 13, 2017. http://www.personalgenomes.org/us.

Pinola, Melanie. "Turn off Windows 8's default sharing of your location, name, and web content usage." *ITWorld*, November 7, 2012. https://www.itworld.com/article/2717530/consumerization/turn-off-windows-8-s-default-sharing-of-your-location--name--and-web-content-usage.html.

ProCon.org. "Drones ProCon.org." *ProCon.org*. May 1, 2017. http://drones.procon.org.

Ponemon Institute. "Aftermath of a Data Breach Study." 2012. http://www.experian.com/assets/data-breach/brochures/ponemon-aftermath-study.pdf.

______. "2016 Cost of Data Breach Study: Global Analysis." 2016. https://www.cloudmask.com/hubfs/IBMstudy.pdf.

Porter, Catherine M. "The Black Death and Persecution of the Jews." *Saber and Scroll* (Winter 2014): 55-65. https://digitalcommons.apus.edu/cgi/viewcontent.cgi?article=1121&context=saberandscroll.

Porup, J.M. "'Internet of Things' security is hilariously broken and getting worse." *ArsTechnica*, January 23, 2016. https://arstechnica.com/information-technology/2016/01/how-to-search-the-internet-of-things-for-photos-of-sleeping-babies.

Postscapes. "IoT Market Forecasts." Accessed July 3, 2017. https://www.postscapes.com/internet-of-things-market-size.

______. "IoT Devices & Products." June 4, 2018. https://www.postscapes.com/internet-of-things-award/winners/.

Presser, Stephen. *Law Professors: Three Centuries of Shaping American Law.* St. Paul: West Academic Publishing, 2016.

Privacy International. "Big Data." Accessed July 3, 2017. https://www.privacyinternational.org/node/8.

______. "Explainers." Accessed July 3, 2017. https://www.privacyinternational.org/privacy-101.

Privacy Rights Clearinghouse. "Data Breaches." Accessed July 4, 2017. https://www.privacyrights.org/data-breaches.

Prosser, William L. "Privacy." *California Law Review* 48.3 (August 1960): 383–423. doi:10.15779/Z383J3C.

Ptashkin, Samantha. "Web Cam of Houston Children's Bedroom Hacked." *Click2Houston*, August 10, 2016. http://www.click2houston.com/news/web-cam-of-houston-childrens-bedroom-hacked.

Ragan, Steve. "Norse Corp Disappears Shortly after CEO Is Asked to Step Down." *CSO from IDG*, February 1, 2016. http://www.csoonline.com/article/3028104/techology-business/norse-corp-disappears-shortly-ceo-is-asked-to-step-down.html.

Rangachari, P K. "Evidence-based medicine: old French wine with a new Canadian label?" *Journal of the Royal Society of Medicine* (May 1997): 280-284. https://www.ncbi.nlm.nih.gov/pmc/articles/PMC1296268.

Ranger, Steve. "Big Data: Why the Boom Is Already Over." *ZDNet*, October 4, 2016. http://www.zdnet.com/article/big-data-why-the-boom-is-already-over.

Rawlinson, Kevin, and Patrick Greenfield. "UK government accused of 'pathetic' lack of help for Anguillans hit by Irma." *The Guardian*, September 6, 2017. https://www.theguardian.com/world/2017/sep/06/government-accused-of-disgraceful-lack-of-aid-for-anguillans-hit-by-irma

Reaves, Jessica. "Anthrax: Separating Fear From Fact." *TIME,* October 12, 2001. http://content.time.com/time/nation/article/0,8599,178989,00.html

Reel, Monte. "Secret Cameras Record Baltimore's Every Move from Above." *Bloomberg,* August 23, 2016. https://www.bloomberg.com/features/2016-baltimore-secret-surveillance.

Regalado, Antonio. "Does Your Genome Predict Your Face? Not Quite Yet." *MIT Technology Review,* September 7, 2017. https://www.technologyreview.com/s/608813/does-your-genome-predict-your-face-not-quite-yet

Resnick, Gideon and Shane Harris. "Will Brexit Masterminds Work for Trump?" *Daily Beast,* July 3, 2016. http://www.thedailybeast.com/will-brexit-masterminds-work-for-trump.

Reuters Staff. "China electronics firm to recall some U.S. products after hacking attack." *Reuters*, October 24, 2016. https://www.reuters.com/article/us-cyber-attacks-manufacturers-idUSKCN12O0MS.

Reynolds, Matthew. "Yahoo's Anti-Abuse AI Can Hunt Out Even the Most Devious Online Trolls." *Wired,* July 29, 2016. http://www.wired.co.uk/article/yahoo-online-abuse-algorithm.

Richards, Neil. *Intellectual Privacy: Rethinking Civil Liberties in the Digital Age.* New York: Oxford University Press, 2015.

Risen, Tom. "FAA Considering Drone Flights Over Crowds." *US News,* April 4, 2016. https://www.usnews.com/news/articles/2016-04-04/faa-committee-recommends-allowing-drone-flights-over-crowds.

Roberts, Jeff John. "Google Faces New Pressure Over So-Called Super-Profiles." *Fortune,* January 25, 2017. http://fortune.com/2017/01/25/google-super-profiles.

Robinson, Bradshaw, and Hinson. "Genomics Law Report." Accessed September 9, 2017. https://www.genomicslawreport.com/index.php/category/general-interest/privacy-general-interest.

Rosen, Ben. "Why a House Bill Wants Workers' Genetic Information." *Christian Science Monitor,* March 12, 2017. https://www.csmonitor.com/USA/2017/0312/Why-a-House-bill-wants-workers-genetic-information.

Rosen, Jeffrey. "The Right to Be Forgotten." *Stanford Law Review Online* 64 (February 2012): 88-92. https://review.law.stanford.edu/wp-content/uploads/sites/3/2012/02/64-SLRO-88.pdf.

Rosenberg, Jonathan B, and Arthur Mateos. *The Cloud at Your Service: The When, How, and Why of Enterprise Cloud Computing.* Greenwich: Manning, 2011.

Roser, Max, Nagdy, Mohamed, and Hannah Ritchie. "Terrorism." *Our World in Data.* Last Revised January 2018. https://ourworldindata.org/terrorism.

Ross, Terrance F. "Welcome to Smarter Basketball." *The Atlantic,* June 25, 2015. https://www.theatlantic.com/entertainment/archive/2015/06/nba-data-analytics/396776.

Rouse, Margaret. "Cloud Computing." Tech Target. Accessed September 9, 2017. http://searchcloudcomputing.techtarget.com/definition/cloud-computing.

Ruff, Tom. "7 reasons why cloud-based security makes sense." *GCN,* March 22, 2017. https://gcn.com/articles/2017/03/22/cloud-based-security.aspx.

Samji, Tasleem. "Influenza A: Understanding the Viral Life Cycle." *Yale Journal of Biology and Medicine* 82 (2009): 153-9. https://www.ncbi.nlm.nih.gov/pmc/articles/PMC2794490/.

Sandberg-Diment, Erik. "The Executive Computer." *The New York Times*, 1985. https://www.nytimes.com/1985/12/08/business/the-executive-computer.html.

Sannappa, Nithan, and Lorrie Cranor. "A Deep Dive into Mobile App Location Privacy Following the InMobi Settlement." Federal Trade Commission, August 9, 2016. https://www.ftc.gov/news-events/blogs/techftc/2016/08/deep-dive-mobile-app-location-privacy-following-inmobi-settlement.

Sarich, Christina. "Pharma Companies Spend 19x more on Marketing than Research, and Returns are Dropping." *Natural Society*, December 21, 2016. http://naturalsociety.com/research-development-new-drugs-not-paying-off-6321.

Sayler, Kelley. "A World of Proliferated Drones: A Technology Primer." Center for a New American Society. Accessed July 1, 2017. http://drones.cnas.org/wp-content/uploads/2016/03/CNAS-World-of-Drones_052115.pdf.

Scherer, Michael. "Inside the Secret World of the Data Crunchers Who Helped Obama Win." *Time,* November 7, 2012. http://swampland.time.com/2012/11/07/inside-the-secret-world-of-quants-and-data-crunchers-who-helped-obama-win.

Schneier, Bruce. *Data and Goliath: The Hidden Battles to Collect Your Data and Control Your World.* New York: W.W. Norton & Company, 2015.

______. "Security of Password Managers." *Schneier on Security,* September 5, 2014. https://www.schneier.com/blog/archives/2014/09/security_of_pas.html.

Schuman, Evan. "Evan Schuman: Starbucks Caught Storing Mobile Passwords in Clear Text." *Computer World,* January 15, 2014. http://www.computerworld.com/article/2487743/security0/evan-schuman--starbucks-caught-storing-mobile-passwords-in-clear-text.html.

Schwartz, Paul M., and Daniel J. Solove. "The PII Problem: Privacy and a New Concept of Personally Identifiable Information." *New York University Law Review* 1814 (December 2011): 1814-94. http://scholarship.law.berkeley.edu/cgi/viewcontent.cgi?article=2638&context=facpubs.

Scott, Mark. "Google Rebuts Antitrust Claims in Europe." *New York Times*, November 3, 2016. https://www.nytimes.com/2016/11/04/technology/google-eu-antitrust-europe.html?mtrref=fortune.com&gwt.

Sealy, P., and M. Menting. *IoT Security and Managed Services.* New York: ABI Research, 2017.

Sekar, Aswin, et al. *Nature.* "Schizophrenia Risk from Complex Variation of Complement Component 4." *Nature* 530 (February 2016): 177–183. doi:10.1038/nature16549.

Serna, Joseph. "Man Allegedly Used Facebook, Instagram to Find Women, Steal Their Underwear." *L.A. Times,* December 8, 2015. http://www.latimes.com/local/lanow/la-me-ln-social-media-undergarment-thief-20151208-story.html.

Sharples, Andrew. "Gene Patents in Europe Relatively Stable Despite Uncertainty in the US." *GenEng News*, March 23, 2011. http://www.genengnews.com/gen-exclusives/gene-patents-in-europe-relatively-stable-despite-uncertainty-in-the-us/77899385.

Sherr, Ian. "WannaCry ransomware: Everything you need to know." *CNET,* May 19, 2017. https://www.cnet.com/news/wannacry-wannacrypt-uiwix-ransomware-everything-you-need-to-know.

Shields, Todd, and Vonnie Quinn. "Facebook Could Be Fined Millions for Violating Consent Deal." *Bloomberg*, March 29, 2018. https://www.bloomberg.com/news/articles/2018-03-29/facebook-risks-millions-of-dollars-in-ftc-fines-over-data-crisis.

ShiSh. "Some Examples of E-commerce Sites hosted on Microsoft Azure." *Microsoft Developer,* May 1, 2015. https://blogs.msdn.microsoft.com/shishirs/2015/05/01/some-examples-of-ecommerce-sites-hosted-on-microsoft-azure.

Shodan. "Shodan." Accessed July 3, 2017. https://www.shodan.io.

Shrier, David, and Alex Pentland. "The Spiraling Arms Race of Data Insecurity." *CNBC*, September 27, 2016. http://www.cnbc.com/2016/09/27/the-spiraling-arms-race-of-data-insecurity-commentary.html.

Singer, P. W., and Allan Friedman. *Cybersecurity and Cyberwar: What Everyone Needs to Know*. Oxford: Oxford University Press, 2014.

Singh, Vivek Kumar, Bozkaya, Burcin, and Alex Pentland. "Money Walks: Implicit Mobility Behavior and Financial Well-Being." PLoS One 10.8 (August 2015). doi:10.1371/journal.pone.0136628.

Sly, Liz. "Al-Qaeda Disavows Any Ties with Radical Islamist ISIS Group in Syria, Iraq." *Washington Post,* February 3, 2014. https://www.washingtonpost.com/world/middle_east/al-qaeda-disavows-any-ties-with-radical-islamist-isis-group-in-syria-iraq/2014/02/03/2c9afc3a-8cef-11e3-98ab-fe5228217bd1_story.html.

Smart Energy GB. "About the Rollout." Accessed July 3, 2017. https://www.smartenergygb.org/en/smart-future/about-the-rollout.

Smolla, Rodney A. *Free Speech in an Open Society*. New York: Knopf, 1992.

_______. *Jerry Falwell v. Larry Flynt: The First Amendment on Trial.* Chicago: University of Illinois Press, 1988.

Snider, Mike. "ISPs can now collect and sell your data: What to know about Internet privacy rules."*USA Today*, April 4, 2017. https://www.usatoday.com/story/tech/news/2017/04/04/isps-can-now-collect-and-sell-your-data-what-know-internet-privacy/100015356.

Springer, Paul J. *Military Robots and Drones: A Reference Handbook*. Santa Barbara: ABC-CLIO, 2013.

Sreeharsha, Vinod. "Google and Yahoo Win Appeal in Argentine Case." *New York Times,* August 19, 2010. http://www.nytimes.com/2010/08/20/technology/internet/20google.html.

Stanek, Becca. "The Childhood Of Dzhokhar Tsarnaev Was Pretty Normal, Leaving Questions As To What Changed." *Bustle*, May 10, 2015. https://www.bustle.com/articles/82291-the-childhood-of-dzhokhar-tsarnaev-was-pretty-normal-leaving-questions-as-to-what-changed.

StatCounter. "Top 5 Search Engines in the United States, Aug. 2016." August 2016. http://gs.statcounter.com/#desktop-search_engine-US-monthly-201608-201608-bar.

Statista. "Number of Smartphone Users in France from 2015 to 2021 (in Millions)." Accessed July 2, 2017. https://www.statista.com/statistics/467177/forecast-of-smartphone-users-in-france.

_______. "Number of Smartphone Users in the United States from 2010 to 2021 (in Millions)." Accessed July 2, 2017. https://www.statista.com/statistics/201182/forecast-of-smartphone-users-in-the-us.

Statt, Nick. "Google will stop scanning your Gmail messages to sell targeted ads." *The Verge*, June 23. https://www.theverge.com/2017/6/23/15862492/google-gmail-advertising-targeting-privacy-cloud-business.

Stieglitz, Jeremy. "Encryption: Pros and Cons." *Imperva Blogs*, September 12, 2017. https://www.imperva.com/blog/2017/09/encryption-pros-and-cons.

Stone, Kathlyn. "Who Funds Biomedical Research?" *The Balance,* April 24, 2017. https://www.thebalance.com/who-funds-biomedical-research-2663193.

Streich, Katja. "Technologies for Server-Based Indoor Positioning Compared: Wi-Fi vs. BLE vs. UWB vs. RFID." infsoft, July 13, 2017. https://www.infsoft.com/blog-en/articleid/188/technologies-for-server-based-indoor-positioning-compared.

Sullivan, Mark. "23andMe Has Signed 12 Other Genetic Data Partnerships beyond Pfizer and Genentech." *Venture Beat*, January 14, 2015. https://venturebeat.com/2015/01/14/23andme-has-signed-12-other-genetic-data-partnerships-beyond-pfizer-and-genentech.

Sweeney, Latanya. "Discrimination in Online Ad Delivery." Cornell University Library (2013). https://arxiv.org/abs/1301.6822.

______. "Simple Demographics Often Identify People Uniquely." Data Privacy Working Paper 3, 2000. https://dataprivacylab.org/projects/identifiability/paper1.pdf.

Swire, Peter, and Jennifer Daskal. "What the CLOUD Act means for privacy pros." *IAPP Privacy Tracker*, March 26, 2018. https://iapp.org/news/a/what-the-cloud-act-means-for-privacy-pros.

Synergy Research Group, "Cloud Growth Rate Increased Again in Q1; Amazon Maintains Market Share Dominance." April 27, 2018. https://www.srgresearch.com/articles/cloud-growth-rate-increased-again-q1-amazon-maintains-market-share-dominance.

Szoldra, Paul. "This Is Everything Edward Snowden Revealed in One Year of Unprecedented Top-Secret Leaks." *Business Insider,* September 16, 2016. http://www.businessinsider.com/snowden-leaks-timeline-2016-9.

Taleb, Nassim N. "Beware the Big Errors of 'Big Data.'" *Wired,* February 8, 2013. https://www.wired.com/2013/02/big-data-means-big-errors-people.

Tanner, Adam. "Different Customers, Different Prices, Thanks to Big Data." *Forbes*, April 14, 2014. https://www.forbes.com/sites/adamtanner/2014/03/26/different-customers-different-prices-thanks-to-big-data.

Taplin, Jonathan. "Is It Time to Break Up Google?" *New York Times,* April 22, 2017. https://www.nytimes.com/2017/04/22/opinion/sunday/is-it-time-to-break-up-google.html.

Taylor, Glenda. "5 Types of Screwdrivers Every DIYer Should Get to Know." *Bob Vila.com.* Accessed June 8, 2018. https://www.bobvila.com/articles/types-of-screwdrivers.

Techonomy. "Business and the Economy in an Internet + Everything World." September 17, 2013. http://techonomy.com/conf/13-detroit/the-new-techonomy/business-and-the-economy-in-an-internet-everything-world.

Telegraph Reporters. "Designer goods worth £400,000 stolen from John Terry's mansion while he was on holiday." *Telegraph,* July 28, 2017. http://www.telegraph.co.uk/news/2017/07/27/designer-goods-worth-400000-stolen-john-terrys-mansion-holiday.

Terlep, Sharon, Vranica, Suzanne, and Shayndi Raice. "GM Says Facebook Ads Don't Pay Off." *Wall Street Journal,* May 16, 2012. https://www.wsj.com/articles/SB10001424052702304192704577406394017764460.

The Economist. "Securing the Digital City: Cyber-Threats and Responses." *NEC,* 2016. http://safecities.economist.com/wp-content/uploads/sites/5/2016/09/NEC_Securing-the-digital-city_final.pdf.

______. "Why the American legal system is so flexible." February 2, 2017. https://www.economist.com/books-and-arts/2017/02/02/why-the-american-legal-system-is-so-flexible.

Thompson, Nicholas. "Instagram Unleashes an AI System to Blast Away Nasty Comments." *Wired,* June 29, 2017. https://www.wired.com/story/instagram-launches-ai-system-to-blast-nasty-comments.

Thompson, II, Richard M. "Drones in Domestic Surveillance Operations: Fourth Amendment Implications and Legislative Responses." Congressional Research Service. Accessed July 1, 2017. https://fas.org/sgp/crs/natsec/R42701.pdf.

Thornton, Mark. "Alcohol Prohibition Was a Failure." *Cato Institute,* July 17, 1991. https://www.cato.org/publications/policy-analysis/alcohol-prohibition-was-failure.

Tiku, Nitasha. "Europe's New Privacy Law Will Change the Web, and More." *Wired,* March 19, 2018. https://www.wired.com/story/europes-new-privacy-law-will-change-the-web-and-more.

Tippmann, Sylvia, and Julia Powles. "Google Accidentally Reveals Data on 'Right to Be Forgotten' Requests." *The Guardian,* July 14, 2015. https://www.theguardian.com/technology/2015/jul/14/google-accidentally-reveals-right-to-be-forgotten-requests.

Toplensky, Rochelle. "EU Fines Google €2.4bn over Abuse of Search Dominance." *Financial Times*, June 27, 2017. https://www.ft.com/content/9554a8bc-5b12-11e7-b553-e2df1b0c3220?mhq5j=e3.

Transparency Market Research. "Cloud Computing Services Market—Global Industry Analysis, Size, Share, Growth, Trends, and Forecast 2016–2024." Accessed July 9, 2017. http://www.transparencymarketresearch.com/cloud-computing-services-market.html.

Trost, Brett, et al. "Comparing the Similarity of Different Groups of Bacteria to the Human Proteome." PLoS One 7.4 (April 2012). doi: 10.1371/journal.pone.0034007.

Tyler Clementi Foundation. "Tyler's Story." Accessed July 7, 2017. https://tylerclementi.org/tylers-story.

Tyson, Peter. "The Hippocratic Oath Today." *NOVA Online*, March 27, 2001. http://www.pbs.org/wgbh/nova/body/hippocratic-oath-today.html.

U.S. Congress. House of Representatives. "FAA Modernization and Reform Act of 2012 Conference Report (to Accompany H.R. 658)." 112th Congress, 2d Session, 2012. H.R. Rep. 112-381. https://www.gpo.gov/fdsys/pkg/CRPT-112hrpt381/pdf/CRPT-112hrpt381.pdf.

______. Senate. Office of Oversight and Investigations Majority Staff, Committee on Commerce, Science, and Transportation. "A Review of the Data Broker Industry: Collection, Use,and Sale of Consumer Data for Marketing Purposes." December 18, 2013. https://www.commerce.senate.gov/public/_cache/files/0d2b3642-6221-4888-a631-08f2f255b577/AE5D72CBE7F44F5BFC846BECE22C875B.12.18.13-senate-commerce-committee-report-on-data-broker-industry.pdf.

U.S. Department of Health and Human Services. "Office for Human Research Protections." Accessed December 13, 2017. https://www.hhs.gov/ohrp.

______. "HIPAA for Professionals." June 16, 2017. https://www.hhs.gov/hipaa/for-professionals/index.html.

U.S. Department of Health and Human Services Office for Civil Rights. "Breach Portal: Notice to the Secretary of HHS Breach of Unsecured Protected Health Information." Accessed July 1, 2017. https://ocrportal.hhs.gov/ocr/breach/breach_report.jsf.

U.S. Food and Drug Administration. "FDA allows marketing of first direct-to-consumer tests that provide genetic risk information for certain conditions." April 6. 2017. https://www.fda.gov/NewsEvents/Newsroom/PressAnnouncements/ucm551185.htm.

______. "Statement from FDA Commissioner Scott Gottlieb, M.D., on implementation of agency's streamlined development and review pathway for consumer tests that evaluate genetic health risks." November 6, 2017. https://www.fda.gov/NewsEvents/Newsroom/PressAnnouncements/ucm583885.htm.

U.S. National Library of Medicine. "Tay-Sachs Disease." Accessed December 13, 2017. https://www.ncbi.nlm.nih.gov/pubmedhealth/PMHT0024672.

U.S. Senate. "The Interstate Commerce Act is Passed." Accessed June 4, 2018. https://www.senate.gov/artandhistory/history/minute/Interstate_Commerce_Act_Is_Passed.htm.

van Engelen, Errol. "Big Data Will Effectively Fight Terrorism in the World." Datafloq.com, January 22, 2017. https://datafloq.com/read/big-data-will-effectively-fight-terrorism/785.

Vanian, Jonathan. "How Bad Polling Data Fooled Everyone Except Donald Trump." *Fortune,* November 10, 2016. http://fortune.com/2016/11/10/donald-trump-data-analytics.

———. "You No Longer Have to Register Your Drone." *Fortune,* May 19, 2017. http://fortune.com/2017/05/19/drone-faa-registration-appeals-hobbyist.

Vasgar, Jeevan. "Berlin Pushes Google to Reveal Search Engine Formula." *Financial Times*, September 15, 2014. https://www.ft.com/content/9615661c-3ce1-11e4-9733-00144feabdc0?mhq5j=e1.

Veterans Administration. "Safeguarding Veterans' Information." US Department of Veteran Affairs. August 26, 2015. https://www.research.va.gov/for_veterans/Safeguarding-Veterans-Info.pdf.

Von Zeipel, Emma. "FAA Issues New Rules for Commercial Drones." *Law Street Media*, June 22, 2016. https://lawstreetmedia.com/blogs/technology-blog/faa-issues-new-rules-commercial-drones-amazon-will-wait.

W3C. "Standards." Accessed July 7, 2017. https://www.w3.org/standards.

Wachter, Sandra, Mittelstadt, Brent, and Luciano Floridi. Why a Right to Explanation of Automated Decision-Making Does Not Exist in the General Data Protection Regulation." *International Data Privacy Law,* 2017. doi: 10.2139/ssrn.2903469.

Waddell, Kaveh. "Why Google Quit China—and Why It's Heading Back." *The Atlantic,* January 19, 2016. https://www.theatlantic.com/technology/archive/2016/01/why-google-quit-china-and-why-its-heading-back/424482.

Waldrop, M. Mitchell. "The chips are down for Moore's Law." *Nature* 530 (February 2016): 144-7. doi: 10.1038/530144a.

Wallace, Nick. "EU's Right to Explanation: A Harmful Restriction on Artificial Intelligence." *TechZone 360,* January 25, 2017. http://www.techzone360.com/topics/techzone/articles/2017/01/25/429101-eus-right-explanation-harmful-restriction-artificial-intelligence.htm.

Wallen, Jack. "Five nightmarish attacks that show the risks of IoT security." *ZDNet,* June 1, 2017. https://www.zdnet.com/article/5-nightmarish-attacks-that-show-the-risks-of-iot-security.

Warner, JH. "Attitudes to Foreign Knowledge." *The Therapeutic Perspective: Medical Practice, Knowledge, and Identity in America*, 1820-35. Cambridge, MA: Harvard University Press, 1986: 105-206. From

Warren, Samuel and Louis Brandeis. "The Right to Privacy." *Harvard Law Review* (1890). doi: 10.2307/1321160.

Washington Post. "How the NSA Is Tracking People Right Now." Accessed July 2, 2017. http://apps.washingtonpost.com/g/page/national/how-the-nsa-is-tracking-people-right-now/634.

Waters, Michael. "The Public Shaming of England's First Umbrella User." *Atlas Obscura,* July 27, 2016. https://www.atlasobscura.com/articles/the-public-shaming-of-englands-first-umbrella-user.

Waters, Richard. "Unrest Over Google's Secret Formula." *Financial Times.* July 11, 2010. https://www.ft.com/content/1a5596c2-8d0f-11df-bad7-00144feab49a?mhq5j=e3.

Watson, Sara M. "Ask the Decoder: Stalked by Socks." *Al Jazeera America,* September 16, 2014. http://america.aljazeera.com/articles/2014/9/16/the-decoder-stalkedbysocks.html.

Wauters, Robin. "Right to Be Found: Google Has Removed 50 Links to Wikipedia Content from Its European Search Results So Far (after 5 Requests)." *Tech.eu*, August 6, 2014. http://tech.eu/news/wikipedia-right-to-be-forgotten-google.

Wein, Berel. "The Black Death." *JewishHistory.org*. Accessed December 13, 2017. https://www.jewishhistory.org/the-black-death.

Weinman, Joe. *Cloudonomics: The Business Value of Cloud Computing.* Hoboken, NJ: John Wiley & Sons, 2012.

Weinstein, Mark. "Why Does Europe Hate Silicon Valley?" *Huffington Post,* June 26, 2015. http://www.huffingtonpost.com/mark-weinstein/why-does-europe-hate-silicon-valley_b_7645172.html.

Wensierski, Peter. "East German Snitching Went Far Beyond the Stasi." *Spiegel Online,* July 10, 2015. http://www.spiegel.de/international/germany/east-german-domestic-surveillance-went-far-beyond-the-stasi-a-1042883.html.

White House. "Presidential Memorandum: Promoting Economic Competitiveness While Safeguarding Privacy, Civil Rights, and Civil Liberties in Domestic Use of Unmanned Aircraft Systems." February 15, 2015. https://obamawhitehouse.archives.gov/the-press-office/2015/02/15/presidential-memorandum-promoting-economic-competitiveness-while-safegua.

______. "Remarks by the President at the National Defense University." May 23, 2013. https://obamawhitehouse.archives.gov/the-press-office/2013/05/23/remarks-president-national-defense-university.

WiFi Map. "WiFi Map." Accessed June 20, 2018. https://www.wifimap.io.

WiGLE. "Frequently Asked Questions." Accessed June 20, 2018. https://wigle.net/faq.

Wilde, Elizabeth Ty, et al. "Do Emergency Medical System Response Times Matter for Health Outcomes?" *Health Economics* 22 (July 2013): 790-806. doi: 10.1002/hec.2851.

Winograd, David. "12 Years a Slave Sees Surge in Illegal Downloads After Oscar Win." *Time*, March 4, 2014. http://entertainment.time.com/2014/03/04/12-years-a-slave-illegal-downloads-bittorrent.

Woolf, Nicky. "DDoS attack that disrupted internet was largest of its kind in history, experts say." *The Guardian,* October 26, 2016. https://www.theguardian.com/technology/2016/oct/26/ddos-attack-dyn-mirai-botnet.

Worden, Scott. "Micropayments and the Future of the Web." Harvard University. Accessed July 3, 2017. https://cyber.harvard.edu/fallsem98/final_papers/Worden.html.

World Health Organization. "Gender and Genetics." Accessed December 13, 2017. http://www.who.int/genomics/gender/en/index1.html.

______. "Plague." October 2017. http://www.who.int/mediacentre/factsheets/fs267/en.

______. "World Immunization Week 2012." Accessed December 13, 2017. http://www.who.int/immunization/newsroom/events/immunization_week/2012/further_information/en.

Worldometers. "US Population (Live)." Accessed July 2, 2017. http://www.worldometers.info/world-population/us-population.

Wu, Tim. "Network Neutrality, Broadband Discrimination." *Journal of Telecommunications and High Technology Law* 2 (2003): 141-179. https://ssrn.com/abstract=388863.

WuXi PharmaTech (Cayman) Inc. "WuXi PharmaTech Acquires NextCODE Health to Create Global Leader in Genomic Medicine." *PR Newswire,* January 9, 2015. https://www.prnewswire.com/news-releases/wuxi-pharmatech-acquires-nextcode-health-to-create-global-leader-in-genomic-medicine-300018311.html.

Yakowitz, Jane. 2011. "Tragedy of the Data Commons." *Harvard Journal of Law & Technology* 25.1 (Fall 2011): 1–67. doi: 10.2139/ssrn.1789749

Young, Jesse. "Best Drone for GoPro [12 Outstanding GoPro Drones 2018]." *Drone Guru,* April 24, 2018. http://www.droneguru.net/best-drone-for-gopro.

ZDNet. "Sports Analytics: How 'Moneyball' Meets Big Data (gallery)." March 14, 2013. http://www.zdnet.com/pictures/sports-analytics-how-moneyball-meets-big-data-gallery.

Zhang, Sarah. "The Loopholes in the Law Prohibiting Genetic Discrimination." *The Atlantic,* March 13, 2017. https://www.theatlantic.com/health/archive/2017/03/genetic-discrimination-law-gina/519216.

Zichur, Kathryn, and Aaron Smith. "4% of Online Americans Use Location-Based Services." *Pew Research Center*, November 4, 2010. http://www.pewinternet.org/2010/11/04/4-of-online-americans-use-location-based-services.

Zipline. "How Zipline Works." Accessed Jan. 12, 2018. http://www.flyzipline.com/service.

About the Author

Achutha Raman, founder of 4PrivacyMatters, is an information privacy consultant who provides advice on privacy strategies, governance, operations, and best practices. He is a Certified Information Privacy Professional (CIPP/U.S.) and graduate of Yale College. Passionate about community outreach to promote privacy for the benefit of consumers, he is eager to engage with anyone and everyone interested. He can be reached at achutha.raman@4privacymatters.org.

Index

94230152R00216

Made in the USA
Lexington, KY
25 July 2018